NUCLEAR MATERIALS AND DISASTER RESEARCH

# SPENT NUCLEAR FUEL

# STORAGE OPTIONS AND EFFORTS

# NUCLEAR MATERIALS AND DISASTER RESEARCH

Additional books in this series can be found on Nova's website under the Series tab.

Additional E-books in this series can be found on Nova's website under the E-book tab.

# ENERGY POLICIES, POLITICS AND PRICES

Additional books in this series can be found on Nova's website under the Series tab.

Additional E-books in this series can be found on Nova's website under the E-book tab.

NUCLEAR MATERIALS AND DISASTER RESEARCH

# SPENT NUCLEAR FUEL STORAGE OPTIONS AND EFFORTS

PHILLIP T. CRAWFORD
AND
ALAN E. MASON
EDITORS

Copyright © 2012 by Nova Science Publishers, Inc.

**All rights reserved.** No part of this book may be reproduced, stored in a retrieval system or transmitted in any form or by any means: electronic, electrostatic, magnetic, tape, mechanical photocopying, recording or otherwise without the written permission of the Publisher.

For permission to use material from this book please contact us:
Telephone 631-231-7269; Fax 631-231-8175
Web Site: http://www.novapublishers.com

### NOTICE TO THE READER

The Publisher has taken reasonable care in the preparation of this book, but makes no expressed or implied warranty of any kind and assumes no responsibility for any errors or omissions. No liability is assumed for incidental or consequential damages in connection with or arising out of information contained in this book. The Publisher shall not be liable for any special, consequential, or exemplary damages resulting, in whole or in part, from the readers' use of, or reliance upon, this material. Any parts of this book based on government reports are so indicated and copyright is claimed for those parts to the extent applicable to compilations of such works.

Independent verification should be sought for any data, advice or recommendations contained in this book. In addition, no responsibility is assumed by the publisher for any injury and/or damage to persons or property arising from any methods, products, instructions, ideas or otherwise contained in this publication.

This publication is designed to provide accurate and authoritative information with regard to the subject matter covered herein. It is sold with the clear understanding that the Publisher is not engaged in rendering legal or any other professional services. If legal or any other expert assistance is required, the services of a competent person should be sought. FROM A DECLARATION OF PARTICIPANTS JOINTLY ADOPTED BY A COMMITTEE OF THE AMERICAN BAR ASSOCIATION AND A COMMITTEE OF PUBLISHERS.

Additional color graphics may be available in the e-book version of this book.

**Library of Congress Cataloging-in-Publication Data**

ISBN 978-1-62257-347-9

*Published by Nova Science Publishers, Inc. † New York*

# CONTENTS

| | | |
|---|---|---|
| **Preface** | | **vii** |
| **Chapter 1** | U.S. Spent Nuclear Fuel Storage<br>*James D. Werner* | **1** |
| **Chapter 2** | Experience Gained from Programs to Manage<br>High-Level Radioactive Waste and Spent Nuclear<br>Fuel in the United States and Other Countries<br>A Report to Congress and the Secretary of Energy<br>*United States Nuclear Waste Technical Review Board* | **73** |
| **Chapter 3** | Spent Fuel Storage in Pools and Dry Casks:<br>Key Points and Questions and Answers<br>*United States Nuclear Regulatory Commission* | **159** |
| **Index** | | **173** |

# PREFACE

Recent events have renewed long-standing congressional interest in safe management of spent nuclear fuel (SNF) and other high level nuclear waste. These issues have been examined and debated for decades, sometimes renewed by world events like the 9/11 terrorist attacks. The incident at the Fukushima Dai-ichi nuclear reactor complex in Japan, combined with the termination of the Yucca Mountain geologic repository project, have contributed to the increased interest. This book focuses on the current situation with spent nuclear fuel storage in the United States. It addresses the SNF storage situation, primarily at current and former reactor facilities and former reactor sites for the potentially foreseeable future. Although no nation has yet established a permanent disposal repository for SNF and other forms of high-level radioactive waste, there is broad consensus that a geological repository is the preferred method for these wastes.

Chapter 1 – Regardless of the outcome of the ongoing debate about the proposed Yucca Mountain geologic waste repository in Nevada, the storage of spent nuclear fuel (SNF)—also referred to as "high-level nuclear waste"—will continue to be needed and the issue will continue to be debated. The need for SNF storage, even after the first repository is opened, will continue for a few reasons.

First, the Obama Administration terminated work on the only planned permanent geologic repository at Yucca Mountain, which was intended to provide a destination for most of the stored SNF. Also, the Yucca Mountain project was not funded by Congress in FY2011 and FY2012, and not included in the Administration's budget request for FY2013.

Second, even if the planned repository had been completed, the quantity of SNF and other high-level waste in storage awaiting final disposal now exceeds the legal limit for the first repository under the Nuclear Waste Policy Act (NWPA).

Third, the expected rate of shipment of SNF to the repository would require decades to remove existing SNF from interim storage. Accordingly, the U.S. Nuclear Regulatory Commission (NRC) and reactor operators are considering extended SNF storage lasting for more than 100 years.

The debate about SNF typically involves where and how it is stored, as well as what strategies and institutions should govern SNF storage. The earthquake and tsunami in Japan, and resulting damage to the Fukushima Dai-ichi nuclear power plant, caused some in Congress and NRC to consider the adequacy of protective measures at U.S. reactors. The NRC Near-Term Task Force on the disaster concluded it has "not identified any issues that undermine our confidence in the continued safety and emergency planning of U.S. plants." Nonetheless, NRC has accepted a number of staff recommendations on near-term safety enhancement, including requirements affecting spent fuel storage and prevention and coping with station blackout. NRC is not requiring accelerated transfer of SNF from wet pools to dry casks, but the SNF storage data from the last several years indicate that accelerated transfer has already been occurring.

As of December 2011, more than 67,000 metric tons of SNF, in more than 174,000 assemblies, is stored at 77 sites (including 4 Department of Energy (DOE) facilities) in the United States located in 35 states (see Table 1 and Figure 5), and increases at a rate of roughly 2,000 metric tons per year.

Approximately 80% of commercial SNF is stored east of the Mississippi River. At 9 commercial SNF storage sites there are no operating nuclear reactors (so-called "stranded" SNF), and at the 4 DOE sites reactor operations largely ceased in the 1980s, but DOE-owned and some commercial SNF continues to be stored at DOE facilities. In the United States, SNF is stored largely at nuclear reactor sites where it was generated.

Of the 104 operating nuclear reactors in the United States, all necessarily have wet storage pools for storing SNF (wet pools are required to allow for a safe "cooling off" period of 1 to 5 years after discharge of SNF from a reactor). Wet storage pools are used for storage of approximately 73% (49,338 out of 67,450 metric tons of uranium, or MTU) of the current commercial SNF inventory, whereas the remaining 27% (18,112 MTU) of commercial SNF is stored in dry casks on concrete pads or in vaults.

As wet storage pools become filled to capacity using "dense packing" storage methods, dry storage is increasingly being used, although there are 27 sites with 36 wet storage pools with no dry cask storage capabilities.

This chapter focuses on the current situation with spent nuclear fuel storage in the United States. It does not address all of the issues associated with permanent disposal of SNF, but rather focuses on the SNF storage situation, primarily at current and former reactor facilities for the potentially foreseeable future.

Chapter 2 – This chapter explores how 13 nations are carrying out efforts to find a permanent solution for isolating and containing high-level radioactive waste (HLW) and spent nuclear fuel (SNF) generated within their borders. Many forces shape how those efforts are designed and implemented. Some of the forces are technical, including choices made about what reactor technology to adopt and about what nuclear fuel cycle to pursue. Others are social and political in nature, including how concerns about intergenerational equity should be addressed and what pace should be followed in implementing a long-term management option. Importantly, the interdependencies, both subtle and overt, between the technical, social, and political forces are inescapable. Because of those interdependencies, what characterizes the national programs most notably is their *variety*. This chapter attempts to detail that variety. It builds on the information contained in the U.S. Nuclear Waste Technical Review Board's (NWTRB) *Survey of National Programs for Managing High-Level Radioactive Waste and Spent Nuclear Fuel* (NWTRB 2009).

Chapter 3 – This chapter contains information from the United States Nuclear Regulatory Commission publication, adapted from the following website: http://www.nrc.gov/ waste/spent-fuel-storage/faqs.html, dated March 2012.

In: Spent Nuclear Fuel        ISBN: 978-1-62257-347-9
Editors: P.T. Crawford and A.E. Mason © 2012 Nova Science Publishers, Inc.

*Chapter 1*

# U.S. SPENT NUCLEAR FUEL STORAGE[*]

## *James D. Werner*

## SUMMARY

Regardless of the outcome of the ongoing debate about the proposed Yucca Mountain geologic waste repository in Nevada, the storage of spent nuclear fuel (SNF)—also referred to as "high-level nuclear waste"—will continue to be needed and the issue will continue to be debated.

The need for SNF storage, even after the first repository is opened, will continue for a few reasons. First, the Obama Administration terminated work on the only planned permanent geologic repository at Yucca Mountain, which was intended to provide a destination for most of the stored SNF. Also, the Yucca Mountain project was not funded by Congress in FY2011 and FY2012, and not included in the Administration's budget request for FY2013. Second, even if the planned repository had been completed, the quantity of SNF and other high-level waste in storage awaiting final disposal now exceeds the legal limit for the first repository under the Nuclear Waste Policy Act (NWPA). Third, the expected rate of shipment of SNF to the repository would require decades to remove existing SNF from interim storage. Accordingly, the U.S. Nuclear Regulatory Commission (NRC) and reactor operators are considering extended SNF storage lasting for more than 100 years.

---

[*] This is an edited, reformatted and augmented version of a Congressional Research Service publication, CRS Report for Congress R42513, , dated May 3, 2012.

The debate about SNF typically involves where and how it is stored, as well as what strategies and institutions should govern SNF storage. The earthquake and tsunami in Japan, and resulting damage to the Fukushima Dai-ichi nuclear power plant, caused some in Congress and NRC to consider the adequacy of protective measures at U.S. reactors. The NRC Near-Term Task Force on the disaster concluded it has "not identified any issues that undermine our confidence in the continued safety and emergency planning of U.S. plants." Nonetheless, NRC has accepted a number of staff recommendations on near-term safety enhancement, including requirements affecting spent fuel storage and prevention and coping with station blackout. NRC is not requiring accelerated transfer of SNF from wet pools to dry casks, but the SNF storage data from the last several years indicate that accelerated transfer has already been occurring.

As of December 2011, more than 67,000 metric tons of SNF, in more than 174,000 assemblies, is stored at 77 sites (including 4 Department of Energy (DOE) facilities) in the United States located in 35 states (see Table 1 and Figure 5), and increases at a rate of roughly 2,000 metric tons per year.

Approximately 80% of commercial SNF is stored east of the Mississippi River. At 9 commercial SNF storage sites there are no operating nuclear reactors (so-called "stranded" SNF), and at the 4 DOE sites reactor operations largely ceased in the 1980s, but DOE-owned and some commercial SNF continues to be stored at DOE facilities. In the United States, SNF is stored largely at nuclear reactor sites where it was generated.

Of the 104 operating nuclear reactors in the United States, all necessarily have wet storage pools for storing SNF (wet pools are required to allow for a safe "cooling off" period of 1 to 5 years after discharge of SNF from a reactor). Wet storage pools are used for storage of approximately 73% (49,338 out of 67,450 metric tons of uranium, or MTU) of the current commercial SNF inventory, whereas the remaining 27% (18,112 MTU) of commercial SNF is stored in dry casks on concrete pads or in vaults.

As wet storage pools become filled to capacity using "dense packing" storage methods, dry storage is increasingly being used, although there are 27 sites with 36 wet storage pools with no dry cask storage capabilities.

This chapter focuses on the current situation with spent nuclear fuel storage in the United States. It does not address all of the issues associated with permanent disposal of SNF, but rather focuses on the SNF storage situation, primarily at current and former reactor facilities for the potentially foreseeable future.

# Introduction

Recent events have renewed long-standing congressional interest in safe management of spent nuclear fuel (SNF) and other high level nuclear waste.[1] These issues have been examined and debated for decades, sometimes renewed by world events like the 9/11 terrorist attacks. The incident at the Fukushima Dai-ichi nuclear reactor complex in Japan, combined with the termination of the Yucca Mountain geologic repository project,[2] have contributed to the increased interest.

This report focuses on the current situation with spent nuclear fuel storage in the United States. It does not address all of the issues associated with permanent disposal of SNF, but rather focuses on the SNF storage situation, primarily at current and former reactor facilities and former reactor sites for the potentially foreseeable future (i.e., a total of 300 years).[3]

Although no nation has yet established a permanent disposal repository for SNF and other forms of high-level radioactive waste (HLW), there is broad consensus that a geological repository is the preferred method for these wastes.[4] In the United States, the disposal repository location has been debated for decades. The proposed repository in Nevada at Yucca Mountain was terminated in 2009, although the project continues to be debated and litigated. Whether the current situation with the Yucca Mountain project is merely a temporary hiatus or becomes a permanent shutdown, extended storage for longer than previously anticipated is virtually assured. There is also no clear consensus on interim storage of SNF. The SNF storage issues most widely debated include:

- What strategies should be employed for SNF storage, pending disposal;
- Where should SNF be stored on an interim basis; and
- What SNF management structure should be used?

A focus of this renewed broader debate is the recently completed report by the Blue Ribbon Commission on America's Nuclear Future (BRC), which released its final report on January 29, 2012.[5]

This final report modified a July 2011 draft, which followed nearly more than a year of effort, including extensive public testimony and several subcommittee reports after the Commission was chartered by the Secretary of Energy in early 2010.[6]

# James D. Werner

---

**Spent Nuclear Fuel (SNF)—In Brief**

- Commercial SNF is comprised of metal assemblies about 12'-15' long (*Figure 1*).
- SNF contains uranium and elements created in nuclear reaction.
- SNF assemblies are removed from reactors after being used to produce power.
- Existing reactors generate about 2,000 metric tons per year.
- More than 67,000 metric tons of commercial SNF is currently being stored
- Most SNF is stored at 77 sites in 35 states (see *Figure 5*).
- Some SNF is stored at closed reactors.
- Some SNF is stored at Department of Energy (DOE) facilities.
- Only 4% of SNF in U.S. is DOE-owned.
- SNF is stored in wet pools and dry casks.
- SNF storage at reactors was intended to be temporary, pending disposal.
- No nation operates a disposal site for SNF.
- Proposed U.S. disposal site for SNF at Yucca Mountain in Nevada was terminated in 2009.
- SNF storage is expected to be needed for more than 100 years.

---

Some, including the U.S. Nuclear Regulatory Commission (NRC), view the current situation as providing adequate safety. Others, including the National Academy of Sciences, have observed that safety may be improved by wider use of dry storage methods. Some analysts, including the BRC, have considered issues beyond safety—including cost and impact on investments—and urge construction of an interim centralized storage at a "volunteer" location after sufficient cooling has occurred.[7]

Some have recommended a more limited approach to consolidation of interim SNF storage using a smaller number of existing operational reactor sites, away from the original generating reactor, for SNF located where there is no operating reactor generating additional SNF or if repository delays result in greater stranded SNF and higher financial liabilities.[8] The two primary technologies being employed in the United States are wet pool storage and dry cask storage (see "How Is Spent Fuel Stored Now?"). Senator Dianne Feinstein, Chair of the Senate Subcommittee on Energy and Water

# U.S. Spent Nuclear Fuel Storage

Development Appropriations, with jurisdiction over NRC and the U.S. Department of Energy (DOE), stated at a hearing in March 2011:

> Most significantly, I believe we must rethink how we manage spent fuel. Spent fuel must remain in pools for at least five to seven years, at which time it can be moved to safer dry cask storage. However, these pools often become de facto long-term storage, with fuel assemblies "re-racked" thus increasing the heat load of the pools. In California, for instance, fuel removed from reactors in 1984 is still cooling in wet spent fuel pools.... Reports out of Japan indicate there were no problems with the dry casks at Dai-ichi. To me, that suggests we should at least consider a policy that would encourage quicker movement of spent fuel to dry cask storage.[9]

Senator Feinstein followed up this hearing statement with a formal letter to the NRC chairman asking NRC "to seriously consider regulatory policies that would encourage the movement of nuclear fuel, once sufficiently cool, out of spent fuel pools and into dry cask storage systems."[10] Citing the 2006 study by the National Academy of Sciences (NAS) National Research Council (*Safety and Security of Commercial Spent Nuclear Fuel Storage*), Senator Feinstein specifically asked the "NRC to initiate a rulemaking process to immediately require a more rapid shift of spent fuel to dry casks."[11] Senator Feinstein indicated her concern about spent fuel management in evaluating proposals for funding small modular reactors.[12] NRC subsequently considered SNF storage needs a part of its Near-Term Task Force on the Fukushima Dai-ichi incident.

A House Appropriations Subcommittee report recently expressed concern about current spent fuel storage, indicating that "[c]onsolidation of this material in a single site that provides enhanced safety and security will improve public comfort with nuclear power, reduce potential safety and security risk, and fulfill the federal government's obligation under the Nuclear Waste Policy Act of 1982 to assume responsibility of spent fuel."[13] Although consolidated interim storage of SNF has received widespread support as a general concept for many years, proposals for SNF storage at specific locations have been vigorously opposed.[14]

The issue of SNF storage is inextricably linked to related longer-term issues like nuclear power plant operations and construction, SNF reprocessing, and establishment of a permanent repository. This report is focused on SNF storage, and does not provide a detailed examination of these related issues.[15]

Worldwide, there are 436 operational nuclear power reactors in 32 countries and 122 permanently shut-down nuclear power reactors, including several in countries where SNF continues to be stored after the reactor fleet has been shut down.[16] No country, including the United States, has yet established an operating permanent disposal site for SNF or other forms of high-level nuclear waste. All nations rely—to varying degrees—on long-term SNF storage.[17] Although the United States has not fully addressed its nuclear waste issues, DOE has, since 1999, operated a permanent geological repository for plutonium-contaminated transuranic (i.e., mainly plutonium-contaminated)[18] waste from nuclear weapons operations[19] in New Mexico. Known as the Waste Isolation Pilot Plant (WIPP), this waste disposal repository was constructed and began operations following decades of detailed technical and institutional planning, active state and community involvement,[20] and compliance work to meet various federal and state environmental review and permitting requirements.[21] The WIPP site is explicitly prohibited by law from receiving SNF or high level waste, but is widely regarded as a model for other waste facility—both disposal and storage—siting efforts. A recent survey of spent fuel storage in the 10 countries with significant nuclear operations found that all countries store substantial amounts of SNF in pools or dry cask facilities,[22] regardless of their policy on reprocessing.[23]

Although much of the congressional attention has focused on the issues of the permanent geological repository proposed for Yucca Mountain in Nevada, there are a number of reasons to also consider SNF storage issues.

First, under any scenario for waste acceptance into a permanent repository or an interim consolidated storage site, long-term storage of SNF will be required for a considerable time. Notwithstanding the mandate in the Nuclear Waste Policy Act (NWPA) and various contracts that DOE begin accepting SNF for disposal in 1998, no disposal repository has been completed or licensed. The 2009 termination of the Yucca Mountain disposal project continues the ongoing delay in opening a permanent geologic repository.[24] Hence, the disposal delay prolongs storage needs.

Even if a disposal repository were to begin operation quickly, the time required to ship SNF would require an extended period of storage. In its most recent estimate, prior to termination of the Yucca Mountain repository program, DOE projected that, if waste acceptance were to begin in 2020, there would be a need for commercial interim storage until at least 2056, given this projected shipment rate and the continued generation of new SNF.[25]

Also, current law—the NWPA—sets a limit on how much waste can be put in the first repository, and the U.S. inventory of SNF and other high level waste requiring disposal has already exceeded this limit. The NWPA "prohibit[s] the emplacement in the first repository of a quantity of spent fuel containing in excess of 70,000 metric tons of uranium (MTU)[26] or a quantity of solidified high-level radioactive waste resulting from the reprocessing of such a quantity of spent fuel until such time as a second repository is in operation."[27] Of this 70,000 MTU limit set by Congress on the first repository in the NWPA, approximately 90% (63,000 MTU) of the capacity is allocated to commercial spent nuclear fuel and high-level radioactive waste from reprocessing.[28] The remaining 10% capacity would be used for about 2,455 MTU of DOE spent nuclear fuel (including naval spent nuclear fuel) and the equivalent of 4,667 MTU of DOE high-level radioactive waste.[29] Hence, the current quantity of SNF (i.e., 67,450 MTU of civilian SNF and 2,458 MTU of DOE-owned SNF) and high-level waste being stored would fill the proposed Yucca Mountain repository beyond the limit imposed by Congress in the NWPA, necessitating a need to build a second repository or change the legal limit. DOE has evaluated the capacity of the proposed Yucca Mountain repository and concluded that Yucca Mountain could hold more than the legal limit of 70,000 MTU and "has the physical capability to allow disposal of a much larger inventory."30

If waste were accepted at a consolidated nuclear waste storage site, rather than a disposal repository, the need for interim storage technologies could continue for a longer period. Hence, the issues related to safe long-term SNF storage—regardless of the current debate about a permanent geologic repository—warrant consideration. A recent study by the Massachusetts Institute of Technology (MIT) concluded that "planning for the long term interim storage of spent nuclear fuel—for about a century—should be part of fuel cycle design."[31] The issues and options associated with current national policies are discussed below.

The recent change in the Yucca Mountain repository program is not the first time that concern about the path forward of a permanent geologic repository has caused increased attention to longterm, possibly consolidated, storage of SNF. After the failure of the salt dome waste repository program near Lyons, KS, the Atomic Energy Commission[32] proposed in 1972 a program for longterm (100-year) retrievable surface storage.[33] In 1987, the same amendments to the Nuclear Waste Policy Act (NWPA) that identified the Yucca Mountain site also established the Monitored Retrievable Storage Commission (ended in 1989) and a position of the "Nuclear Waste

Negotiator," which was eliminated in 1995 after several years of unsuccessful attempts to find a voluntary host community for a repository or monitored retrievable storage facility for nuclear waste.[34] During the 1990s, Congress tried, unsuccessfully, to enact legislation to help establish a temporary consolidated storage site for commercial spent nuclear fuel near the Yucca Mountain site. A number of analysts have long recommended that interim storage of spent fuel be implemented deliberately for a period of at least 100 years.[35]

Second, SNF stored outside of a reactor comprises a source of radioactivity requiring durable protection. Although radioactive decay reduces the amount of radioactivity in SNF, dropping sharply soon after discharge from a reactor, SNF provides a significant and long-term radioactivity source term for risk analyses. While the nuclear fuel in an operating reactor typically contains a larger amount of radioactivity, in terms of curie content, than stored SNF, much of this curie content is composed of relatively short-lived fission products, and includes more volatile constituents, compared to SNF. Although there have been some releases of radioactivity from stored SNF (see "Hazards and Potential Risks Associated with SNF Storage" below), there is no evidence of any consequent significant public exposures or health impacts. Nonetheless, significant concerns have been raised about the potential for releases from stored SNF.[36] These concerns have been heightened in the wake of the incident at the Fukushima Dai-ichi reactors in Japan, about which there have been conflicting accounts and some uncertainty regarding the condition of the stored SNF.

Extended storage has also raised concerns about long-term site safety. For example, in the wake of general concern about the risks from extreme weather and sea level rise from climate change, as well as specific concerns about SNF stored near flood-prone rivers (e.g., along the Missouri River), some have expressed urgency about the need to relocate SNF storage.[37] NRC, however, found that sea level rise was not a credible threat to existing and planned on-site nuclear waste storage for the next several decades: "Based on the models discussed in the [National Academy of Sciences/National Research Council study], none of the U.S. [Nuclear Power Plants] (operational or decommissioned) will be under water or threatened by water levels by 2050."[38]

Third, the federal government faces a significant and growing liability to pay claims resulting from its failure to begin accepting waste from commercial utilities under the 1987 NWPA.[39] The U.S. government[40] has paid approximately $1 billion[41] to pay a series of claims by utilities that DOE had, at least partially, breached its contracts to accept SNF.[42] The federal government has been paying claims for commercial utility costs for SNF storage since 2000.[43] These claims arise from the 76 standard contracts DOE signed in 1983, largely with commercial utilities, of which 74 have filed claims against DOE for damages arising from failure to accept the SNF by 1998.[44]

The future estimated costs for storage of commercial SNF are approximately $500 million per year.[45]

Fourth, some have argued that the uncertainty and concerns about nuclear waste management have contributed to the lack of investment in new nuclear power plants, resulting in a failure of the industry to expand, along with relatively high capital costs.[46] An American Physical Society Panel chaired by a former NRC chairman, and including another former NRC chairman and a former Under Secretary of Energy, concluded, in part, "there is a concern that the buildup of spent fuel at reactor sites and lack of progress on final disposition could be serious constraints on the growth of the domestic nuclear power industry by discouraging investment in new nuclear power plants and enhancing the difficulty of siting new nuclear power plants."[47]

A recent article by long-time nuclear waste observers and former officials argued that "solid public acceptance of nuclear energy ... may well turn on a credible promise of a geologic repository becoming available within the next few decades."[48] Another longtime observer of the nuclear industry indicated, "[e]ven if the public were otherwise prepared to go along with a major expansion of nuclear power, much less reprocessing, it is unlikely to do so without a new, credible regime for disposing of our existing and future nuclear power wastes."[49]

The recent report by the BRC also implicated the current "impasse" in the U.S. nuclear waste program as a hindrance to expansion of nuclear power, among other impacts:

> Put simply, this nation's failure to come to grips with the nuclear waste issue has already proved damaging and costly and it will be more damaging and more costly the longer it continues: damaging to prospects for maintaining a potentially important energy supply option for the future, damaging to state-federal relations and public confidence in the federal government's competence, and damaging to America's standing in the world—not only as a source of nuclear technology and policy expertise but as a leader on global issues of nuclear safety, non-proliferation, and security.[50]

States have imposed their own controls on SNF management and related nuclear power plant operations.[51] Specifically, some state laws prohibit construction of any new nuclear power plants until the current backlog of spent fuel is addressed. According to the National Conference of State Legislatures, "Thirteen states have laws prohibiting energy utilities from even considering adding new nuclear reactors until the waste problem has been solved."[52] State authority, however, is limited under the Atomic Energy Act.[53]

This report cannot resolve these issues, but it does provide some vital background to help support an informed debate on the issues of SNF storage.

## WHAT IS SPENT (USED) FUEL?

In many cases, discussions of "nuclear waste"[54] are, in fact, referring to SNF. The "fuel" in commercial nuclear fuel is uranium oxide[55] formed into solid cylindrical ceramic pellets, contained in zirconium alloy tubes supported in a rigid metal framework, or "assembly" (see *Figure 1*). These fuel assemblies are composed of individual rods of approximately half an inch in diameter and 12 to 15 feet long, with each assembly approximately 5 inches to 9 inches on a side.[56] The fuel assemblies for a boiling water reactor (BWR) are about half the mass (about 0.18 MTU/assembly) of a typical pressurized water reactor (PWR), which are about 0.44 MTU/assembly.[57] The difference in design of BWR and PWR reactors is significant in how SNF is stored, which is discussed below. Both types of reactors are "light water reactors" because they use ordinary ("light") water for cooling and reducing neutron energy.[58] All current U.S. commercial reactors are light water reactors.

The production of electricity through nuclear fission in a light water reactor uses low-enriched uranium (3%-5% U-235) pellets that undergo a nuclear fission process inside the reactor. The fission process heats water to generate steam that turns the turbine generator, thereby generating electricity. In general, the fuel rods are productive for approximately 54 months. Roughly every 18 months, utilities generally conduct a refueling outage in which approximately one-third of the fuel assemblies are replaced with new assemblies.

A replaced fuel assembly becomes spent (or "used") nuclear fuel when it has been irradiated and removed from a nuclear reactor after it is no longer cost-effective to generate power. This reduced power output is caused by the accumulation of radionuclides generated by the fission (splitting) of the uranium atoms and the relative reduction in fissile isotopes.[59]

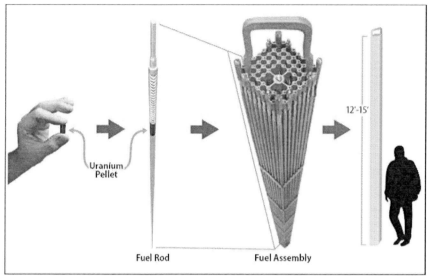

Source: CRS-produced graphic using SNF assembly image from General Electric Company.

Note: Illustrates a typical light water reactor nuclear fuel assembly used in U.S. commercial nuclear power reactors, and may not resemble fuel for research reactors, naval nuclear propulsion reactors, or nuclear power reactor fuel in other countries.

Figure 1. Nuclear Fuel Assembly.

These fission products effectively "poison" the nuclear chain reaction by interfering with the otherwise self-sustaining fission process. The SNF assemblies must be removed to maintain the necessary power production level to generate electricity.

A typical commercial nuclear power plant fuel rod assembly operates in a reactor for approximately 4 1/2 years before being removed for storage and eventual disposal.[60] Once removed from the reactor core, SNF continues to generate heat and radiation, and requires careful management for thousands of years. Commercial reactors in the United States generate about 2,000 MT of SNF annually.[61]

The variety of SNF types is as diverse as the range of reactors and their functions, which is dominated by, but not limited to, commercial nuclear power plants designed to produce electric power. This report is focused on commercial SNF, in contrast to smaller quantities of other types of SNF, which include SNF with different designs and characteristics from other sources:

- DOE production reactors used for producing nuclear materials, such as Pu-239 (weapons-grade plutonium) and Pu-238 (for deep space missions and other applications requiring durable power sources);[62]
- reactors used for research, analysis of materials, basic science experiments, and training;[63] and
- naval propulsion reactors used for submarines and aircraft carriers.[64] The Naval Nuclear Propulsion Program (i.e., nuclear submarines, aircraft carriers, and Navy prototype and training reactors) has generated about 27 MTU of SNF, which is stored at the Idaho National Laboratory.[65]

Virtually all of this SNF is DOE-owned and comprises less than 4% of the amount of commercial SNF stored in the United States (2,458 MTU vs. 67,450 MTU). These noncommercial SNF types include a wide range of designs and sizes, which differ significantly from commercial reactor fuel, but share the same basic radioactivity hazard and need for long-term isolation. Many of the technical issues associated with long-term storage of commercial SNF also apply to these fuel types. This inventory of DOE-owned SNF does not include the millions of gallons of liquid high-level waste stored in underground tanks or the canisters of vitrified high-level waste resulting from reprocessing SNF at DOE sites.

## HOW IS SPENT FUEL STORED NOW?

There are essentially two modes of SNF storage in the United States: wet pools and dry casks.[66] Wet storage pools are the most common method for storing SNF in the United States, accounting for about 73% (49,338 MTU) of the current commercial SNF inventory.The remaining 27% (18,112 MTU) of commercial SNF is stored in dry casks on concrete pads or horizontal bunkers. Operating commercial nuclear power plants must store SNF at least a year (often five years or more) in "spent fuel pools" to allow for some initial cooling after discharge from the reactor. Because of limited wet storage capacity for SNF, most sites employ a combination of wet pool and dry cask storage.

NRC regulates both wet fuel pools[67] and dry cask installations[68] and regards both as adequately protective (see "Options for Storing Spent Nuclear Fuel" below).

# U.S. Spent Nuclear Fuel Storage

Wet storage pools are constructed of reinforced concrete walls several feet thick, with stainless steel liners. The water is typically about 40 feet deep, and serves both to shield workers from radiation and cool the SNF assemblies. Storage pools vary somewhat in size, but are generally large enough to store fuel rods vertically, with ample depth to provide space above the SNF storage racks for unloading and loading SNF transfer casks with SNF that is approximately 12 to 15 feet long. The cooling water chemistry is generally carefully controlled to minimize corrosion. Just as the fuel assemblies' designs differ among different reactor types, described above, the designs of the pool storage basins differ significantly among different reactor designs, often unique to each plant. Currently, commercial nuclear reactors in the United States are light water reactors, of which there are two basic designs: boiling water reactors (BWRs; see *Figure 2*), and pressurized water reactors (PWRs; see *Figure 3*), the former being the type of reactor built at the Fukushima Dai-ichi site in Japan. Some SNF being stored in the United States, however, was generated by other reactor types[69] not currently used in this country, such as high temperature gas-cooled reactors and liquid sodium reactors.[70]

The design of storage pools has been highlighted recently for the GE Mark I BWR, which has been used at dozens of nuclear power plants worldwide, including the Fukushima Dai-ichi plant, as well as 20 reactors operating in the United States. The SNF storage pool for the GE Mark I reactor is approximately 35 feet wide, 40 feet long, and 39 feet deep (10.7 meters wide, 12.2 meters long, and 11.9 meters deep), with a water capacity of almost 400,000 gallons (1.51 million liters). Another feature of the GE Mark I BWR is that the SNF storage pool is located inside the same secondary containment structure as the reactor and many critical control systems. Finally, the SNF storage pools in the BWR Mark I reactors are located several stories above ground level. The potential safety considerations of these design features are discussed below in "Hazards and Potential Risks Associated with SNF Storage."The SNF storage capacities using wet storage pools at U.S. commercial power reactors generally range from approximately 2,000 assemblies to 5,000 assemblies (averaging approximately 3,000 SNF fuel assemblies). Typically, U.S. spent fuel pools are filled with spent fuel assemblies up to approximately three-quarters of their capacity to allow space for at least one full reactor core load of fuel to be stored as needed. In contrast, the wet storage pool building located nearby but separate from the reactors at the Fukushima Dai-ichi site contains about 6,000 spent fuel assemblies, which was about half of the SNF at the site. U.S. reactor facilities do not typically have an additional spent fuel wet storage building for on-site SNF consolidation like that at Fukushima Dai-ichi.[71]

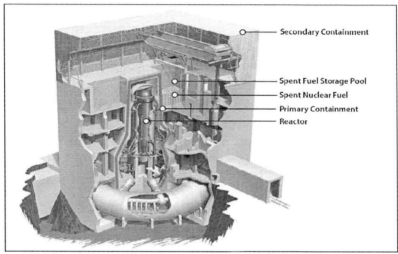

Source: GE-Hitachi Nuclear Energy.
Note: Illustrates a GE Mark I BWR design, which is one of a number of BWR designs.

Figure 2. SNF Storage Pool Location in Boiling Water Reactor.

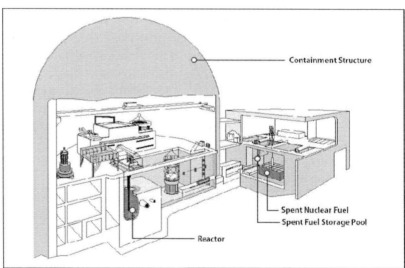

Source: Timothy Guzda, Union of Concerned Scientists, modified by CRS.

Figure 3. PWR SNF Storage Pool Location.

A critical feature of wet pools is the need for power to provide makeup water and circulate the water to keep it from boiling off and uncovering the fuel, especially for recently discharged SNF that is still thermally and radioactively hot. The SNF loses much of its heat in the first few weeks after discharge, and after five years produces relatively little thermal energy. Nonetheless, without circulation pumps helping remove the accumulated heat, the heat emitted from SNF can be sufficient to boil away the cooling water.

Dry casks are typically constructed in a cylindrical shape with an inner steel canister directly storing the SNF assemblies that is bolted or welded closed, and an outer concrete cask (see *Figure 4*). After loading with SNF, dry casks are stored outside[72] vertically on a purpose-built concrete pad, or horizontally in a storage bunker. An individual SNF storage cask can weigh more than 100 MT (220,000 pounds) and be more than 15 feet long and 6 feet outside diameter.[73] One regularly used SNF storage container (the "NUHOMS 61BT") weighs 22 tons empty and 44 tons loaded with SNF.[74] The largest dry casks licensed for use in the United States can hold up to 40 PWR spent fuel assemblies or 68 BWR spent fuel assemblies. The Transportation, Aging, and Disposal (TAD) Canister System proposed by DOE can hold up to 21 PWR or 44 BWR spent fuel assemblies.[75] The storage capacity of a dry cask depends not only on size but also on the burn-up and age of the SNF fuel to be stored. There are more than 50 different types of dry casks produced by about a dozen manufacturers approved by NRC for general use in the United States.[76] NRC regulates dry cask storage systems at 10 C.F.R. 72, and with various guidance documents (e.g., NUREG-1536).

The key feature of dry storage units is that, once constructed, filled, and sealed, they require no power for circulation of cooling water and are generally regarded as "passively safe." Natural convection of air through the outer concrete shell of the storage casks is sufficient to cool the steel containment cask inside without reliance on power for pumps or fans. By contrast, a wet storage pool system typically requires power for recirculation and/or makeup of cooling water. Loss of power for a wet pool storage system for more than a few days or weeks could cause water levels to drop to levels resulting in potential risks (e.g., inadequate shielding or exposure of SNF).[77] The robustness of dry cask storage was illustrated recently by the results of an August 2011 earthquake centered in Mineral, VA, near the North Anna nuclear power plant.[78] According to the NRC, the earthquake caused movement of the SNF dry storage casks, weighing more than 100 tons, of approximately 1 to 41/2 inches, and had no significant impact on the casks or the SNF.

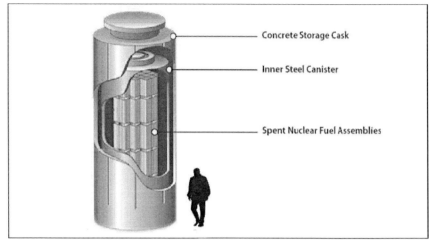

Source: U.S. Nuclear Regulatory Commission.

Note: This figure illustrates a free-standing upright design. Other dry cask designs include horizontal and in-ground vaults.

Figure 4. Example of Dry Cask Storage.

### International Nuclear Waste Convention

The United States is a party[79] to the Joint Convention on the Safety of Spent Fuel Management and on the Safety of Radioactive Waste Management.[80] The waste treaty includes provisions covering transboundary waste shipments, and facility planning, operations and closure, as well as requiring annual reports by each "Contracting Party." This treaty does not create any new obligations on U.S. SNF management, and provides a framework for a variety of strategies employed by a number of countries.

NRC issues both general licenses and site-specific licenses for dry storage facilities.[81] A general license allows licensees to avoid repeating certain evaluations (e.g., National Environmental Policy Act or seismic review), if they have already been conducted pursuant to the plant operating licensing process. Part 72 contains NRC's regulations for the dry storage of power reactor spent fuel on or off a reactor site and for pool storage away from a reactor site.

The NRC regulations have been adopted consistent with the International Convention on Nuclear Waste (see inset box).

The first commercial dry SNF storage system in the United States was established at the North Anna nuclear power plant site in Virginia in 1986, and employs both vertical casks and horizontal modules.[82] Since then, an increasing number of storage locations have employed dry casks at an increasingly rapid rate. The amount of SNF transferred to dry cask storage in 2010 (8,606 assemblies) was nearly four times the average amount transferred since the use of dry cask storage began in 1986 (2,309 assemblies/year),[83] and was more than 50% greater than the assemblies transferred to dry casks in 2009.[84] The amount of SNF stored in wet pools compared to dry systems dropped from 91% (44,712 MTU/49,401 MTU) in 2002 to 73% (49,338 MTU/67,450 MTU) in 2011.

The amount of SNF stored in wet pools increased by about 10%, from 44,712 MTU in 2002 to 49,338 in December 2011, at an average annual rate of 437 MTU/year (1% average annual increase). The amount of SNF stored in dry systems increased during the same period by 286%, from 4,689 MTU to 18,112 MTU. The rate change in the amount of stored SNF reflected an average annual increase of total stored SNF of 1,579 MTU/year (3.2% average annual increase) and average increase in dry system storage of 1,124 MTU/year (24%/year average annual increase)—that is, storage of SNF in dry systems increased, on average, 24 times faster than storage in wet pools.

Regardless of government policies and requirements about SNF storage, it is likely that the trend toward more dry cask storage will continue because of the need for storage and the changing economic value, and extended pay-off period, of these investments given the long-term uncertainty about SNF disposal.

## WHERE IS SPENT NUCLEAR FUEL LOCATED NOW IN THE UNITED STATES?

Spent nuclear fuel is stored at 77 different sites[85] in the United States, including 63 sites with licensed operating commercial nuclear power reactors, 4 DOE-operated sites, 9 former operating nuclear reactor sites and the Morris, IL, proposed reprocessing plant. Generally, the storage sites include facilities at the 104 licensed operating nuclear power reactor locations where it was generated (see *Table 1* and *Figure 5*), as well as 10 "stranded" commercial sites where no reactors operate (including the Morris, IL, proposed reprocessing site), and 4 DOE-operated facilities. In fact, virtually every site that has ever hosted a commercial nuclear reactor is currently also a storage site for SNF. As discussed above, because of practical, technical, and logistical limitations and other issues, SNF is likely to stay there for a significant period, regardless of decisions on a permanent geologic repository. NRC has proposed plans that would consider SNF storage for up to 300 years, in the case of the oldest SNF now in storage.[86]

Of the 67,450 MTU[87] of commercial SNF stored in the United States, approximately three-quarters of it (49,338 MTU) is being stored in wet pools at the reactor sites. Of the 74 different sites[88] (some with multiple storage facilities) where commercial spent fuel is stored in the United States, 55 locations employ dry cask storage for at least part of the storage for commercial SNF.[89] There are 27 sites with 36 wet pool storage "facilities" (i.e., in some cases there are multiple wet pool storage "facilities" co-located at individual sites) where wet pool storage is the only technology being used at the site.[90]

SNF from commercial nuclear power reactors is currently stored at 73 commercial sites around the United States. Sixty-three SNF storage sites also have operating commercial nuclear power reactors. At another 10 commercial sites[91] and an additional four DOE-operated sites,[92] SNF is being stored where there is no operating reactor. At these "stranded" SNF storage sites, the nuclear reactors that generated the SNF have been shut down and at least partly decommissioned.[93] Virtually none of the SNF at these reactors has been moved from where it was generated to another site. In some cases, where reactors have been shut down and decommissioned, the SNF has been moved to another reactor site for storage. Most SNF storage is located in or near the operating nuclear reactor (or sister reactors) that originally generated the SNF (see *Table 1*).

The SNF storage at nine commercial sites where reactors have been shut down has warranted special attention by some in the nuclear industry, plant operators, utilities, and public service commissions monitoring costs. These stranded sites represent roughly half of the sites where reactors have shut down, but where SNF continues to be stored.[94] Because these stranded sites do not share overhead costs (e.g., security, maintenance and utilities) with a larger operating reactor complex, the incremental storage costs are higher than at operating reactor sites.

## Table 1. U.S. Spent Nuclear Fuel Storage Inventories by State (Ranked by Total SNF Mass) As of December 31, 2011 (see Table A-1 for data sorted by states alphabetically)

| State (15 states have no stored SNF)a | Number of Facilities | Number of Sites | "Stranded" SNF Storage Sitesb | Mass (metric tons of uranium) | | | | Assemblies | | |
|---|---|---|---|---|---|---|---|---|---|---|
| | | | | Wet Storage | Dry Cask | Wet Total SNF | Storage | Dry Cask | Total SNF | Table Notes |
| Illinois | 15 | 8 | 2 | 6,900 | 1,791 | 8,691 | 28,242 | 9,625 | 37,867 | b & c |
| Pennsylvania | 9 | 5 | - | 4,606 | 1,459 | 6,065 | 20,898 | 8,424 | 29,322 | d |
| South Carolina | 8 | 5 | 1 | 2,236 | 1,808 | 4,044 | 5,001 | 3,896 | 8,897 | b & e |
| New York | 7 | 4 | - | 3,082 | 495 | 3,577 | 12,466 | 1,820 | 14,286 | b & f |
| North Carolina | 5 | 3 | - | 3,018 | 544 | 3,562 | 10,612 | 1,480 | 12,092 | |
| Alabama | 5 | 2 | - | 2,647 | 540 | 3,187 | 10,978 | 2,180 | 13,158 | |
| Florida | 5 | 3 | - | 2,511 | 445 | 2,956 | 5,859 | 1,024 | 6,883 | f |
| California | 7 | 4 | 2 | 2,017 | 916 | 2,933 | 4,750 | 2,486 | 7,236 | b & g |
| Georgia | 4 | 2 | - | 2,018 | 592 | 2,610 | 7,366 | 3,264 | 10,630 | |
| Michigan | 5 | 4 | 1 | 2,058 | 502 | 2,560 | 6,495 | 1,537 | 8,032 | b & f |
| New Jersey | 4 | 2 | - | 2,025 | 529 | 2,554 | 7,489 | 2,535 | 10,024 | h |
| Virginia | 4 | 2 | - | 970 | 1,477 | 2,447 | 2,120 | 3,229 | 5,349 | f |
| Texas | 4 | 2 | - | 2,121 | 0 | 2,121 | 4,522 | - | 4,522 | |
| Connecticut | 4 | 2 | 1 | 1,439 | 613 | 2,052 | 5,050 | 1,467 | 6,517 | b & g |
| Arizona | 3 | 1 | - | 1,052 | 903 | 1,955 | 2,490 | 2,136 | 4,626 | |
| Tennessee | 3 | 2 | - | 1,095 | 470 | 1,565 | 2,386 | 1,024 | 3,410 | |
| Maryland | 2 | 1 | - | 531 | 808 | 1,339 | 1,197 | 1,824 | 3,021 | f |
| Wisconsin | 4 | 3 | 1 | 915 | 419 | 1,334 | 2,603 | 1,088 | 3,691 | b & f |
| Arkansas | 2 | 1 | - | 607 | 726 | 1,333 | 1,336 | 1,600 | 2,936 | |
| Louisiana | 2 | 2 | - | 1,014 | 235 | 1,249 | 3,861 | 1,148 | 5,009 | |
| Minnesota | 3 | 2 | - | 678 | 525 | 1,203 | 2,645 | 1,770 | 4,415 | g |
| Ohio | 2 | 2 | - | 1,083 | 34 | 1,117 | 4,542 | 72 | 4,614 | |
| Nebraska | 2 | 2 | - | 650 | 203 | 853 | 2,825 | 808 | 3,633 | g |

## Table 1. (Continued)

| State (15 states have no stored SNF)a | Number of Facilities | Number of Sites | "Stranded" SNF Storage Sitesb | Mass (metric tons of uranium) | | | | Assemblies | | | Table Notes |
|---|---|---|---|---|---|---|---|---|---|---|---|
| | | | | Wet Storage | Dry Cask | Wet Total SNF | Storage | Dry Cask | Total SNF | | |
| Mississippi | 1 | 1 | - | 602 | 203 | 805 | 3,428 | 1,156 | 4,584 | | |
| Missouri | 1 | 1 | - | 679 | 0 | 679 | 1,696 | - | 1,696 | | |
| Massachusetts | 2 | 2 | 1 | 542 | 122 | 664 | 3,082 | 533 | 3,615 | b | |
| Washington | 2 | 2 | 1 | 319 | 339 | 658 | 1,715 | 1,836 | 3,551 | b | |
| Kansas | 1 | 1 | - | 646 | 0 | 646 | 1,434 | - | 1,434 | | |
| Vermont | 1 | 1 | - | 513 | 111 | 624 | 2,815 | 612 | 3,427 | | |
| New Hampshire | 1 | 1 | - | 455 | 93 | 548 | 944 | 192 | 1,136 | | |
| Maine | 1 | 1 | 1 | 0 | 542 | 542 | - | 1,438 | 1,438 | b | |
| Iowa | 1 | 1 | - | 259 | 217 | 476 | 1,452 | 1,220 | 2,672 | | |
| Oregon | 1 | 1 | 1 | 0 | 345 | 345 | - | 801 | 801 | b | |
| Idaho | 1 | 1 | 1 | 50 | 81 | 131 | 144 | 177 | 321 | b | |
| Colorado | 1 | 1 | 1 | 0 | 15 | 15 | - | 2,208 | 2,208 | b & f | |
| U.S. Total Commercial Site Storage | 119 | 74 | 10 | 46,733 | 15,859 | 62,592 | 165,583 | 56,493 | 222,076 | | |
| U.S. Total DOE Site Storage | 4 | 4 | 4 | 2,605 | 2,243 | 4,848 | 6,860 | 8,117 | 14,977 | | |

Source: The primary source for these data is the Nuclear Energy Institute (NEI) report, "2011 Used Fuel Data," prepared by Gutherman Technical Associates, January 14, 2012. Site-specific data on sites with no operating reactors ("Storage-only Sites") is derived largely from DOE, Report to Congress on the Demonstration of the Interim Storage of Spent Nuclear Fuel from Decommissioned Nuclear Power Reactor Sites, DOE/RW-0596, December 2008. Data for DOE sites were generally from Frank Marcinowski,

Overview of DOE's Spent Nuclear Fuel & High Level Waste; Presentation to the Blue Ribbon Commission on America's Nuclear Future, U.S. DOE, March 25, 2010.

[a]There are currently 15 states with no commercial SNF storage (there may be temporary and relatively small-scale storage of SNF from non-power generating research and academic reactors): Alaska, Delaware, Hawaii, Indiana, Kentucky, Montana, Nevada, New Mexico, North Dakota, Oklahoma, Rhode Island, South Dakota, Utah, West Virginia, Wyoming.

[b] "Stranded" is generally used to refer to SNF stored where the nuclear reactor that generated the SNF has ceased operating and been decommissioned, and the SNF remains at the site. In some cases the wet storage pools have been dismantled and the SNF is stored in dry casks. In the case of the Morris, IL, site, the "stranded"SNF was shipped from other reactor sites for a proposed reprocessing facility that never operated, and no reactor has ever operated at the site. The number of "stranded" SNF storage sites here does not include sites where SNF from decommissioned reactors is stored at sites co-located with operating reactors (e.g., San Onofre (CA), Dresden 1 (IL), or Indian Point (NY)). This table includes U.S. DOE facilities in Colorado, Idaho, South Carolina, and Washington that store commercial SNF, but where reactors have ceased operating, largely in the 1980s.

[c] Includes the Morris, IL, site, operated by GE-Hitachi, which never hosted an operating reactor or generated any SNF. The site was built to serve as an SNF reprocessing plant for which the SNF from other sites was shipped. The facility never operated, and the SNF has remained stored at the site. Many sources categorize the Morris site with dry cask storage sites, because they are all considered "Independent Spent Fuel Storage Installations" (ISFSI), although the Morris Site differs from other ISFSIs because it uses wet pool storage and is categorized accordingly here.

[d] Excludes SNF and debris generated at the Three Mile Island-2 facility removed after the 1979 incident and shipped to DOE's Idaho National Engineering Laboratory (now referred to as "Idaho National Laboratory") .

[e] Includes 29 MTU of SNF, fragments, and nuclear materials stored at the Savannah River Site (SRS) near Aiken, which has been operated by DOE for nuclear weapons material production. In addition to material generated there, the SRS is used to store SNF shipped from commercial reactors (e.g., Carolinas-Virginia Tube Reactor) and from foreign and domestic research reactors using U.S.-origin fuel.

[f] Does not include SNF shipped to DOE federal facilities: 8 MTU from Florida, 1 MTU from Maryland, 12 MTU from Michigan, 16 MTU from New York, 22 MTU from Virginia, and 4 MTU from Wisconsin, as well as part of the SNF from the Fort St. Vrain site in Colorado, which was shipped to Idaho.

[g] Does not include SNF shipped to the Morris, IL, facility (see (b) above) from 4 states: 100 MTU from California, 34 MTU from Connecticut, 185 MTU from Minnesota, and 191 MTU from Nebraska. Includes 132 MTU shipped from other facilities in Illinois to the Morris facility.

[h] The adjacent Salem 1 and 2 and Hope Creek reactors share dry cask storage facilities and could be considered three facilities on a single contiguous site.

At several other commercial sites (e.g., Millstone 1, CT; Dresden 1, IL; Indian Point 1, NY; and San Onofre, CA), SNF is stored at facilities where a reactor has ceased operating but other reactors at the same site continue operating.[95]

In addition, SNF is stored at three DOE-owned sites where the reactors ceased operating.[96]

At one commercial site in Colorado (Fort St. Vrain), the DOE operates an SNF storage facility from a reactor that shut down in 1989. More than 90% of the stranded commercial SNF is located at five sites. Where SNF remains stored in dry cask storage but no wet storage pool exists, there is some concern that this could make it difficult to repackage the SNF if the need arises because SNF transfers are generally done under water.

The Blue Ribbon Commission (BRC) concluded that the need to address this stranded SNF was one of "several compelling reasons to move as quickly as possible to develop safe, consolidated storage capacity on a regional or national basis," which it argued was "[t]he fundamental policy question for spent fuel storage."[97]

*Figure 5* shows generally that most of the SNF storage in the United States is located in the Midwest and on the East Coast. Specifically, measured by mass, more than 80% (54,092 MTU vs. 13,358 MTU) of SNF is stored at sites east of the Mississippi River. Measured by number of SNF assemblies, however, nearly 84% (197,002 assemblies) of the SNF is stored in eastern sites compared to approximately 17% (40,051 assemblies) stored in western sites.

The fact that BWR fuel assemblies are, individually, about half the mass of PWR fuel assemblies, combined with the fact that there are a disproportionate number of BWRs in the East compared with the overall number nationally,[98] helps explain why there is a greater share of SNF counted by assemblies, compared to the SNF mass (MTU), in the East.

Measured by mass of SNF, there is a slightly higher percentage (83%) of SNF stored in wet pools (40,260 out of 54,092 MTU) at eastern sites compared to the national share of SNF in wet storage (73%, or 49,338 out of 67,450 MTU). The corollary is that there is a somewhat disproportionate amount of SNF stored in dry casks at western sites (32%, or 4,280 out of 13,358 MTU) compared to the national distribution of about 27%.

Measured by number of assemblies, the disproportionate number of assemblies (86%, or 172,443 out of 237,053 assemblies) stored in wet pools in the East is greater than the share of wet pool storage at the western sites (15%, or 25,009 out of 40,051 assemblies), and greater than the national distribution (73%, or 172,443 out of 237,053 assemblies).

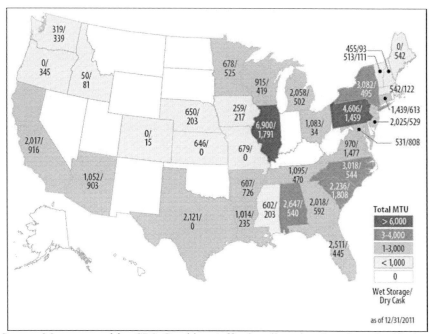

Source: Map prepared by CRS Graphics staff primarily using data from the Nuclear Energy Institute report, "2011 Used Fuel Data," by Gutherman Technical Associates, January 14, 2012. See Table 1 for more detail.

Notes: This map generally reflects storage quantities and location of commercial SNF, although relatively small quantities of commercial SNF is stored at U.S. DOE facilities, The map does not include SNF stored at U.S. DOE sites, which is a small portion (less than 4%) of total stored SNF.

Figure 5. Spent Nuclear Fuel Storage Map.

It is not clear what accounts for this disproportionate share of SNF in dry cask storage at the western sites other than operator preference. Given the smaller size of BWR elements and the disproportionate share of BWRs located at eastern sites, however, it is predictable that the same disproportionate share of SNF by mass is amplified when measured by number of assemblies.

Measured by metric tons of heavy metal content (MTU), the five *states* with the largest *total* amount of SNF stored are

- Illinois (8,691 MTU),
- Pennsylvania (6,065 MTU),
- South Carolina (4,044 MTU; 4,073MTU with DOE SNF),

# U.S. Spent Nuclear Fuel Storage

- New York (3,577 MTU),[99] and
- North Carolina (3,562 MTU).

The top five *states* with the largest amount (by MTU) of SNF stored in *wet pools* are

- Illinois (6,900 MTU),
- Pennsylvania (4,606 MTU),
- New York (3,082 MTU),
- North Carolina (3,018 MTU), and
- Alabama (2,647 MTU).

The top five *states* with the largest amount (by MTU) of SNF in *independent spent fuel storage installations* (generally dry casks, except for the Morris, IL, site) are

- South Carolina (1,808 MTU),
- Illinois (1,791 MTU),
- Virginia (1,477 MTU),
- Pennsylvania (1,49 MTU), and
- California (916 MTU).

The rankings change somewhat if measured by the number of SNF assemblies depending on the portion of different types of reactors in each state (i.e., states with more BWRs have more assemblies per MTU than states with more PWRs).

Measured by number of SNF assemblies, the five *states* with the largest *total* amount of SNF stored are

- Illinois (37,867 assemblies),
- Pennsylvania (29,322 assemblies),
- New York (14,286 assemblies),
- Alabama (13,158 assemblies), and
- North Carolina (12,092 assemblies).

The top five *states* with the largest amount (by assemblies) of SNF stored in *wet pools* are

- Illinois (28,242 assemblies),
- Pennsylvania (20,898 assemblies),
- New York (12,466 assemblies),
- Alabama (10,978 assemblies), and
- North Carolina (10,612 assemblies).

The top five *states* with the largest amount (by assemblies) of SNF in *independent spent fuel storage installations* (generally dry casks, except for the Morris, IL, site) are

- Illinois (9,625 assemblies),
- Pennsylvania (8,424 assemblies),
- South Carolina (3,896 assemblies),
- Georgia (3,264 assemblies), and
- Virginia (3,229 assemblies).

Generally, the total amount of SNF in storage increases by approximately 2,000 MTU per year, assuming current operation of about 104 commercial power reactors. The amount of SNF stored in dry casks versus wet pools reflects more year-to-year changes depending on decisions made by individual plant operators.

Historically, DOE and its predecessor agencies[100] accepted for storage and reprocessing significant amounts of spent nuclear fuel from commercial nuclear power plants. For example, DOE took possession of the spent fuel and debris from the 1979 Three Mile Island plant incident, and shipped it to a DOE facility in Idaho[101] as part of a "research and development" project. In addition, SNF was shipped to New York (West Valley) and Illinois (Morris), where the Atomic Energy Commission, a predecessor to DOE, was involved in reprocessing efforts. In the 1970s a relatively small amount (248.7 MTU) of commercial SNF was shipped from commercial reactors, including utilities in Michigan and New York, to the West Valley site in New York,[102] which reprocessed SNF for about six years (1966 to 1972). The resulting high-level waste and contaminated facilities remain at the site. DOE has estimated that decommissioning and environmental remediation of the contamination at the West Valley site will continue until at least 2020, cost $3.7 billion, and require indefinite long-term stewardship thereafter.[103] The DOE sites in South Carolina and Idaho have also accepted relatively small amounts of SNF from research reactor SNF, some of which contains high-enriched uranium. This report does not attempt to estimate the storage capacity of the existing facilities or sites. To do so would require judgments and determinations about the potential for and safety of re-racking SNF in more dense configurations that are beyond the scope of the analysis.

## SNF MANAGEMENT BY THE U.S. DEPARTMENT OF ENERGY

The U.S. Department of Energy (DOE) has direct responsibility for storage of SNF and other reactor-irradiated nuclear materials at four sites: Fort St. Vrain (CO), Idaho National Laboratory (ID), Savannah River Site (SC), and Hanford Reservation (WA).[104] Because much of the DOE inventory includes material that is fundamentally different from commercial SNF (e.g., enrichment level/U-235 concentration, cladding, size, condition), it is not appropriate to list and sum these materials with commercial SNF. Accordingly, the DOE-owned inventory at these sites is identified distinctly in *Table 1* and *Table A-1* with site numbers "D1" through "D4."

A detailed description of this DOE-owned inventory is beyond the scope of this report, which is focused on commercial SNF stored at commercial sites. A brief summary, however, helps explain why it is accounted for separately in analyzing storage status issues and options. DOE and its predecessor agencies operated a number of nuclear reactors for weapons material production[105] and research. Although DOE ceased operating its large production reactors in the late 1980s and early 1990s, it continues to produce and accept SNF from research reactors. Much of the SNF from these reactors was reprocessed to extract certain nuclear materials (e.g., plutonium), which generated liquid high-level waste, which is intended, after processing, to be disposed of in a permanent geological repository with SNF. Some of DOE's SNF has not been reprocessed and remains stored as solid SNF.

In the case of the Savannah River Site (SRS), for example, there is a wide range of irradiated nuclear materials, including aluminum-based nuclear fuel, "higher actinide targets," and non- aluminum based fuel that DOE has characterized as "assemblies/items."[106] Like the SRS, the SNF at DOE's Hanford site was largely surplus material from reactors used to produce materials (e.g., plutonium) for nuclear warheads, which was extracted using reprocessing facilities at the sites. The design of these fuel elements intended to be reprocessed for weapons material was different from commercial fuel elements (e.g., thin cladding). The Idaho inventory includes SNF from the Naval Nuclear Propulsion program (i.e., submarines and aircraft carriers), which is different from commercial SNF in multiple ways, including enrichment level and design. From about 1952 to 1992 this Navy SNF was reprocessed in Idaho to extract high-enriched uranium for use in driver rods at weapons material production reactors elsewhere.[107] The Idaho site is also home to the SNF debris from the partial meltdown of the Three Mile Island (PA) reactor in 1979. The DOE sites also store a variety of research reactor SNF—both foreign and domestic—that is often much smaller (e.g., 3 to 4 feet long vs. 12 to 15 feet long for commercial power reactor fuel) and more highly enriched than commercial reactor SNF. Some SNF stored at DOE sites came from commercially related activities, including commercial power reactors (e.g., Three Mile Island),[108] test reactors used for academic research (e.g., at 24 U.S. universities),[109] medical isotopes, and commercial power reactor research.[110] By the late 1980s, some of the SNF at DOE sites had become severely corroded or was stored in technically inadequate conditions. After 1993, DOE undertook a "Spent Fuels Vulnerabilities Assessment" and developed detailed "Materials in Inventory" plans to secure this material through packaging and processing, which have been partially implemented generally in conformance with recommendations of the Defense Nuclear Facilities Safety Board.

DOE-owned spent fuel includes some production reactor fuel, which has a fundamentally different design and construction from commercial spent fuel. This DOE-owned production reactor fuel generally has very thin cladding intended to facilitate reprocessing to recover plutonium or other nuclear materials. DOE, with recommendations of the Defense Nuclear Facilities Safety Board, has had a long-term plan to stabilize the relatively thin-walled DOE-owned SNF because it is less robust than commercial SNF, which has thicker and more corrosion-resistant cladding.[111]

The total amount of DOE-owned SNF and other reactor-irradiated material is about 198 MT, of which about half is stored in wet storage pools and the other half in various dry storage configurations. All of the SNF at the Hanford and Fort St. Vrain sites has been transferred to dry casks, while all of the material at the SRS is stored in wet storage pools. The Idaho site employs both methods.

In addition to the different types of SNF and other reactor-irradiated material at DOE sites, DOE is also responsible for high-level waste (HLW) resulting from reprocessing of SNF. Hence, it could be misleading to simply sum the SNF at the DOE-owned sites with the SNF stored at commercial sites, possibly implying a total inventory of HLW intended for extended storage and eventual repository disposal. The SNF inventories at DOE sites must be accounted for separately from commercial sites to compile a meaningful assessment of the total inventory of HLW in the United States. At the Hanford site, the HLW remains stored in underground tanks (many of which have leaked into the ground) awaiting treatment. At the West Valley site in upstate New York, the HLW from SNF reprocessing has largely been vitrified into a borosolicate glass inside stainless steel canisters and awaits disposal. Some of the HLW at the SRS has been similarly treated on a larger scale. At the Idaho sites, the HLW from reprocessing of SNF is being stored in both a liquid form and a solid calcine form awaiting treatment to prepare for disposal. The details of HLW storage, treatment facility construction and operations, "waste incidental to reprocessing," glass log disposition, tank closure, and costs are beyond the scope of this report.[112]

Moreover, the exact inventory of SNF and irradiated materials at DOE sites is more variable than commercial SNF. In some cases DOE may use the SRS reprocessing "H-canyon," for which DOE is required by law to "maintain a high state of readiness"[113] to reprocess materials for recovery of nuclear materials or to stabilize materials for safety reasons.[114] DOE recently announced its intention to use the H-canyon to process plutonium non-pit scrap.[115] Although the exact inventory is difficult to compare because of qualitative differences, such as the non-pit scrap, it is useful to put in context the total SNF inventory at DOE sites to the overall totals. The 29 MTU of SNF at the Savannah River Site comprise about 1.2% of the DOE-owned SNF, and about 0.04% of the total SNF stored in the United States. After reprocessing, the SNF would be converted to a relatively small amount of fissile material, and a larger volume of liquid acidic radioactive waste containing much of the fission products. This liquid radioactive waste would be pumped into underground tanks and (minus any "Waste Incidental to Reprocessing")116 ultimately processed into borosilicate glass logs inside stainless steel canisters destined for disposal in a permanent geological repository. The path forward for this material is beyond the scope of this report.

## HAZARDS AND POTENTIAL RISKS ASSOCIATED WITH SNF STORAGE

Evaluating hazards and risks accurately and with sufficient precision is essential to making decisions about SNF management and regulation. The primary near-term hazard from SNF derives from the radioactivity from the decay of mixed fission products (e.g., cesium-137, strontium-90), and long-term hazards from plutonium and uranium. Hence, spent nuclear fuel generally poses a significant hazard but it may not pose a significant risk, depending on how it is managed. The distinction between "hazard" and "risk" is fundamental to analyzing the need for and benefits of various SNF management options. In classic risk assessment and toxicology, a hazard is the inherent potential of something to cause harm, whereas a risk is generally the product of a probability and the severity of an event (e.g., health, environmental, or financial impact). Whether a hazardous substance poses a risk depends on a variety of factors, including containment and exposure. For example, to the extent that spent fuel is effectively contained, the hazard can remain extremely high, but the risk may be low because there is no pathway for exposure.[117]

Regulation of SNF storage requires a consideration of risk, which involves characterizing the probability and consequence of potential threats. Regulation also requires a policy judgment about what level of assurance is warranted. Some would argue that the hazard, or consequence from the event, is so high that it demands a commensurately high level of protection, using any available technology, against threats to prevent any significant risks to human health and the environment.[118] Others, including a former NRC commissioner, have argued that the mandate of NRC is to "provide reasonable assurance of adequate protection, not absolute assurance of perfect protection."[119]

Although the thermal heat and radiation from SNF begins to drop as soon as the fission process ceases upon reactor shutdown, SNF assemblies continue to require cooling and shielding during storage after discharge from a reactor. Generally, SNF requires three to five years of wet pool storage before it has cooled enough for transfer to dry cask storage. While NRC has authorized transfer as early as three years, the industry norm is about 10 years. Some analysts and observers have expressed concern about the safety of current spent nuclear fuel management practices, at least for long-term storage methods.120 Until the recently proposed termination of the Yucca Mountain

U.S. Spent Nuclear Fuel Storage 31

repository project, however, these SNF storage methods were not generally considered "long-term."

NRC and others have viewed SNF storage as a relatively safe operation, if reasonable precautions are exercised and normal conditions prevail. The inherent hazards of SNF, however, can result in a variety of risks under other conditions. A variety of forces or "threats" acting on spent fuel could result in containment being breached, resulting in potential exposures and risks, generally: (1) loss of power for water supply, circulation, or cooling, which can have significant consequences for SNF in wet pool storage;[121] (2) external threats, like hydrogen explosions from adjacent reactors, or an airplane crashing onto an SNF storage facility; (3) long-term degradation of SNF through chronic corrosion of cladding (e.g., hydride corrosion); and (4) leakage of contaminated water from wet pools to groundwater.

After the earthquake and tsunami in Japan, the first two of these concerns appeared to be occurring at the Fukushima Dai-ichi facility. In testimony on March 16, 2011, before a joint subcommittee hearing of the House Energy and Commerce Committee, NRC chairman Greg Jaczko stated that, following the hydrogen explosion at reactor #4 at the Fukushima Dai-ichi plant, there was an uncovering of the fuel in the pool and there was "no water in the pool."[122] A November 2011 report by the Institute for Nuclear Power Operations concluded that "[S]ubsequent analyses and inspections determined that the spent fuel pool water levels never dropped below the top of fuel in any spent fuel pool and that no significant fuel damage had occurred. Current investigation results indicate that any potential fuel damage may have been caused by debris from the reactor building explosions."[123]

Less than a week after the earthquake and tsunami, and two days after this testimony, NRC published a nonregulatory information notice to reactor operators to "review the information for applicability to their facilities and consider actions, as appropriate, to avoid similar problems."[124] This information notice reiterated NRC's process for adopting earlier Interim Compensatory Measures,[125] which dealt with spent fuel storage safety, among other issues. This initial NRC notice summarized the circumstances, indicating that "Units 3 and 4 have low spent fuel pool (SFP) water levels."[126]

A recent study by the Massachusetts Institute of Technology (MIT), soon after the Fukushima Dai-ichi incident, recommended that NRC and the nuclear power industry

> reexamine safety systems, operating procedures, regulatory oversight, emergency response plans, design basis threats, and spent fuel

management protocols for operating U.S. reactors. Some of these issues were addressed in the aftermath of the TMI-2 accident and the September 11 World Trade Center attacks, resulting both in hardening of U.S. nuclear plants against a number of accident scenarios and in improved emergency response preparations.[127]

The Government Accountability Office has recommended DOE "[a]ssess the condition of existing storage facilities and identify any gaps and actions that might be needed to address longterm storage requirements."[128]

In the wake of the Fukushima Dai-ichi incident, NRC dispatched inspectors to each U.S. reactor and SNF storage site. The results of these inspections varied with different sites, with NRC inspections revealing no significant vulnerabilities for most sites. For sites with the oldest SNF (e.g., at the GE-Hitachi Morris storage site in Illinois), the predicted impact of an extended loss of power to the site would be minimal because of the extent of the radioactive decay during the decades of storage. NRC predicted that, "in the unlikely event that the [spent fuel wet storage basin] is completely drained of water, fuel melt would not occur given the limited fuel decay heat load."[129]

The Fukushima Dai-ichi incident raised concerns about possible risks to SNF storage, including physical damage and loss of power. Although details remain unclear, the dramatic gas explosions that caused some of the overhead containment structure to collapse down into the pool appeared to have caused little or no damage to the SNF in the pools.[130] Also, despite the loss of power over an extended period and some loss of cooling water, the SNF in the storage pools appeared to have remained covered in water largely undamaged.

Notwithstanding these initial results on the SNF condition at the Fukushima Dai-ichi site, some analysts are concerned that, under some circumstances, SNF zirconium cladding can catch fire if left uncovered in wet storage pools. The National Academies/National Research Council concluded, based primarily on review of NRC studies,[131] "that a loss-of-pool-coolant event could trigger a zirconium cladding fire in the exposed spent fuel," but considered "such an accident ...

so unlikely that no specific action was warranted."[132] NRC has issued a series of orders and letters to licensees to implement additional mitigation measures to address the issue.[133] NRC and others, including the nuclear industry, believe these actions provide reasonable assurance of adequate protection. Others disagree, including a long-time nuclear critic who has argued that the large inventory of radioactive material in wet pools poses a significant risk if the SNF pools were drained of water, the SNF were uncovered, and there was a release from an incident such as a fire or deliberate attack.[134]

The locations of SNF wet pool storage in relation to the associated nuclear reactor may present potential risks associated with those designs. For example, most boiling water reactors (BWRs) in the United States, including the GE Mark I, are designed with the SNF storage pool located inside the same secondary containment structure as the reactor and many critical control systems, and located well above ground level. Many have expressed concern that this design may pose safety risks because any problems with the reactor can affect the SNF storage pools, and vice versa.[135] For example, in a loss of off-site power situation, such as occurred at the GE Mark I reactors in Fukushima, Japan, the SNF pool may also lose power, affecting the cooling water and monitoring systems. In the case of the incident in Japan, elevated radiation near the reactor hindered personnel from mitigating problems or monitoring the SNF storage pools. In addition, the height of the SNF pools in many BWRs (more than 100 feet above ground level) could also pose safety risks because of the elevated access challenge and potential for a loss of coolant in a structural failure, compared to reactors with the SNF storage pools at or below ground level.

Prior to the Fukushima Dai-ichi incident, the biggest change in the risk profile for SNF storage occurred in the wake of the September 11, 2001, terrorist attacks, after which a congressionally mandated National Academy of Sciences report concluded that "attacks with civilian aircraft remain a credible threat."[136] NAS indicated that terrorists might choose to attack spent nuclear fuel pools because they are "less well protected structurally than reactor cores" and "typically contain inventories of medium- and long-lived radionuclides that are several times greater than those contained in reactor cores."[137] In response, NRC issued a series of orders and letters to licensees, the contents of which are confidential. NRC also conducted site-specific evaluations to review individual site risks and readiness, resulting in site modifications, the details of which are also confidential. Although the reviews, orders, and letters resulted in numerous incremental improvements to SNF storage facilities and operations, such as improved backup power supply reliability, there was no large-scale shift of SNF out of wet pools and into dry casks, nor was there a mandate to move SNF into hardened storage facilities.

By contrast, Germany explicitly requires protection against risks, including "external events" such as an attack on SNF storage,[138] and this has resulted in construction of hardened storage buildings for dry cask storage of SNF.[139] The Germans have also moved to establish consolidated central SNF storage facilities in the wake of a 2005 decision to cease reprocessing, and the lack of a geological repository.[140]

Another potential threat to SNF storage safety is degradation of the cladding[141] and fuel elements. The potential for degradation of SNF cladding has been well known for decades, and the water chemistry of SNF storage pools is carefully controlled in part to protect against it.[142] The potential vulnerability of SNF varies with the particular type of fuel rod. Generally, however, the rate of corrosion and embrittlement of typical U.S. light water reactor zircalloy-clad commercial fuel rods is less than the rate for the British Magnox fuel rods, or DOE nuclear materials production reactor fuel elements, which are intended for shorter periods inside the reactor.[143] Hence, although some have expressed concern that U.S. nuclear fuel was not designed for long-term storage,[144] U.S. power reactor fuel has generally proven more durable than other fuel forms.

NRC concluded in its initial 1984 waste confidence rule "that the cladding which encases spent fuel is highly resistant to failure under pool storage conditions," and that "[c]orrosion studies of irradiated fuel at 20 reactor pools in the United States suggest that there is no detectable degradation of zircalloy cladding."[145] Nonetheless, BRC and NAS, among others, have recommended long-term research projects to evaluate the integrity of stored SNF,[146] such as ongoing work by the Electric Power Research Institute (EPRI).[147]

NRC has concluded that dry cask storage provides additional confidence in long-term storage of SNF:

> As long as the canister or cask has been properly dried and inerted, and the fuel temperature is kept within the limits of [NRC] guidance, there should be no active degradation mechanisms. The dryness is assured by following the proper drying procedure. No monitoring of the dryness is conducted. The canister is backfilled with an inert gas such as helium. The integrity of the closure seals is actively monitored. If there is indication that the closure seal is compromised, remedial action is taken. The issue of cladding integrity is being reevaluated as part of our extended storage and transportation review and all potential mechanisms for cladding and fuel degradation, even those analyzed to be inactive in the short term, are being reanalyzed for the potential to be active in the longer term.[148]

Finally, NRC has identified releases of tritium-contaminated water to groundwater at 38 sites,[149] and determined that, in some cases,[150] SNF storage pools had contributed[151] to groundwater contamination.[152] In addition to these commercial sites, tritium contamination was found in groundwater from spent fuel storage pools at DOE sites, including the Brookhaven National Laboratory in New York, Hanford in Washington State, and the Savannah River Site in South Carolina.[153] Evidence of groundwater contamination from leaking spent fuel storage pools was among the concerns cited by New York State in its objection to NRC's December 2010 Waste Confidence Rule. Tritium is inherently difficult to remediate, once released, because it is simply a radioactive form of hydrogen that substitutes freely with hydrogen in water and decays at a rate of about 5% per year (12.32 year half life). To address this issue, NRC in March 2010 established a Tritium Contamination Task Force. Notwithstanding these documented releases of radioactivity from stored SNF, there is no evidence of any resulting significant public exposures or health impacts.

Nonetheless, concerns about contamination of groundwater from leaking SNF storage pools were among the issues raised by petitioners challenging NRC's "Waste Confidence Rule." In the wake of these findings, and NRC's own reviews, NRC has issued instructions to licensees and established releases to groundwater as a significant performance indicator, in addition to industry guidance on release prevention and reporting.[154] NRC has argued that these issues presented no public health or environmental risks, and that subsequent "Regulatory Guides," together with inspections, review of license applications, and industry forums and initiatives, will help prevent recurrence.[155] The litigation is ongoing.

NRC's "Waste Confidence Rule" arose from an effort to address the variety of concerns about spent fuel storage in a systematic and explicit manner, and to prevent them from being raised on a site-by-site basis as an issue during individual reactor licensing proceedings. Partly in response to a legal challenge and subsequent remand,[156] NRC issued in 1984 the first of three "Waste Confidence Decisions"[157] that included a series of specific findings, indicating generally that SNF could be stored safely at least until an expected geologic repository began operation to accept wastes. In the first two waste confidence rules, NRC identified a specific date by which a permanent repository would be available and until which SNF could be safely stored.[158] NRC's most recent revision of the rule, however, "removed a date when a repository would be expected to be available for long-term disposal of spent fuel."[159] NRC indicated that SNF could be stored for at least 60 years beyond the end of the reactor operating life:[160]

The [NRC] has made a generic determination that, if necessary, spent fuel generated in any reactor can be stored safely and without significant environmental impacts for at least 60 years beyond the licensed life for operation (which may include the term of a revised or renewed license) of that reactor in a combination of storage in its spent fuel storage basin and at either onsite or offsite independent spent fuel storage installations.[161]

The nuclear power industry has indicated it "is confident that existing dry cask storage technology, coupled with aging management programs already in place, is sufficient to sustain dry cask storage for at least 100 years at reactors and central interim storage."[162]

Aside from the general assurance by NRC about the safety of the SNF storage, one of the implications of the waste confidence determination is

no discussion of any environmental impact of spent fuel storage in reactor facility storage pools or independent spent fuel storage installations (ISFSI) for the period following the term of the reactor operating license or amendment, reactor combined license or amendment, or initial ISFSI license or amendment for which application is made, is required in any environmental report, environmental impact statement, environmental assessment, or other analysis prepared in connection with the issuance or amendment of an operating license for a nuclear power reactor.[163]

The states of New York, Vermont, and Connecticut (later joined by New Jersey) petitioned in February 2011 to challenge NRC's rule, while NEI and Entergy intervened in support of NRC's rule.[164] NRC has indicated plans to consider the need for updating to the Waste Confidence Rule in 2019,[165] after preparing an Environmental Impact Statement.[166] Recognizing the elevated role of extended SNF storage, NRC has shifted resources for FY2012 to its "Spent Fuel Storage and Transportation Business Line" to "evaluate extended long-term storage of radioactive material," including plans for a generic environmental impact statement.[167]

In the wake of DOE's termination of the Yucca Mountain repository program, NRC is undertaking a number of efforts related to the safety of extended SNF storage. U.S. NRC Chairman Greg Jaczko stated:

I have no doubt that we are up to the challenge of addressing the significant policy issues ahead of us. One such issue concerns our approach towards regulating interim and extended spent fuel storage. As part of our Waste Confidence decision, the Commission initiated a comprehensive review of this regulatory framework. This multi-year effort will (1) identify near-term regulatory improvements to current licensing, inspection, and enforcement programs; (2) enhance the technical and regulatory basis for extended storage and transportation; and (3) identify long-term policy changes needed to ensure safe extended storage and transportation. As the question of permanent disposal is for the Congress or the courts to decide, the Commission has been clear that it was neither assuming nor endorsing indefinite, onsite storage by ordering these actions.[168]

Specifically, NRC is undertaking an analysis, as part of its review of the waste confidence rule and the associated Environmental Impact Statement, as well as planning for "Extended Storage and Transportation," that will include a time frame for SNF storage until approximately 2250, which would comprise a total storage period (wet and dry) of approximately 300 years for the oldest SNF now in storage.[169]

## OPTIONS FOR STORING SPENT NUCLEAR FUEL

The Administration's FY2012 budget request included no funding for Yucca Mountain repository operations or licensing activities. The House-passed Energy and Water Development Appropriations bill for FY2012 proposed to restore funding to the Yucca Mountain project, with $25 million for DOE and $20 million for NRC. The final FY2012 appropriations, however, did not include funding for Yucca Mountain activities.[170] The Administration's FY2013 request again includes no funding the Yucca Mountain repository, but instead proposes a variety of general repository studies and fuel cycle research. The recent budget shifts at DOE and NRC will likely be analyzed and debated for years to come. Moreover, the recent incident at the Fukushima Daiichi reactor complex in Japan has spurred reexamination of SNF storage options. Accordingly, reviewing the possible options for SNF storage seems prudent.

Although a complete understanding of the events and their causes at the Fukushima Dai-ichi reactors and spent fuel pools remains incomplete, identification of some basic similarities and differences with U.S. spent fuel storage is possible. NRC established a task force to examine and report to the commission within 90 days on "the agency's regulatory requirements, programs, processes, and implementation in light of information from the Fukushima Dai-ichi site in Japan."[171] This Near-Term Task Force delivered its

90-day report and recommendations to the commission in July 2011,[172] which outlined 12 recommendations for consideration by the commission. They include a number of specific recommendations related to SNF storage safety, largely focused on backup power reliability and monitoring capabilities.[173] The NRC staff identified seven recommendations for near-term action, and offered this overall assessment: "To date the Task Force has not identified any issues that undermine our confidence in the continued safety and emergency planning of U.S. plants ... Task Force review is likely to recommend actions to enhance safety and preparedness."[174]

Options for SNF storage include not only *how* and *where* it is stored, but also what *management and oversight structure* is used. Each of these three issues is addressed below.

A threshold question, however, is what should be the basis for determining the scope of reasonable options for SNF storage? Identifying the reasonable options, and selecting from among them, will depend on certain assumptions and policy choices. Perhaps most importantly, the options and choices depend on the length of time until a permanent geologic repository is available. For decades, virtually every policy analysis has drawn its conclusions and recommendations based on certain assumptions about the opening date for the disposal repository.[175] If that date is delayed indefinitely, it opens many of these earlier assumptions, analyses, and conclusions to reevaluation. As discussed above, uncertainty about the prospects for a permanent disposal repository highlights the need for long-term SNF storage. This report does not make any projections about the timing or path forward for the proposed repository.

Several other assumptions and variables could affect the options for SNF storage, including the amount of SNF generated as a result of reactor license renewal (or denial), the number and frequency of new reactor starts, potential impacts of climate change (sea level rise and more frequent and extreme weather events), or new technical developments; such factors are beyond the scope this report but warrant consideration in a more comprehensive analysis.

## Options for *How* to Store SNF

To consider *how* SNF is stored, there are essentially two options available: wet storage pools and dry cask. Few debate that dry cask storage provides greater safety than wet storage pools. The questions on which there are diverse views include (1) whether wet storage pools provide "adequate" safety, (2) whether the added safety of dry casks is worth the added short-term costs and the potential safety risks during the transfer process, and (3) whether either technology provides adequate safety under extended storage periods (more than 100 years) and under previously unforeseen occurrences.

To some extent, a simple comparison of the two SNF technologies is not appropriate because they are intended to perform somewhat different functions. Wet storage pools provide certain capabilities that dry casks cannot, such as radiological shielding and cooling necessary for intensely radioactive SNF immediately after discharge from reactors. Dry casks, on the other hand, provide adequate shielding and cooling for SNF that has been discharged from reactors at least one to five years. In addition, dry cask storage includes SNF preparation that is a step toward transportation to a repository or a consolidated storage facility (e.g., drying, inert gas, and enclosure in a cask that is typically designed to be transportable) that provides a link in the steps toward final permanent geologic repository disposal.

Notwithstanding this widespread confidence in dry cask storage, this linkage from storage to disposal, however, remains potentially incomplete. One of the significant challenges to managing SNF storage using dry casks is the lingering uncertainty about the approval for the transportation canister. Most of the SNF currently stored in dry casks is contained within an inner multipurpose container that is expected to be used to ship the SNF to a repository and be an essential barrier to prevent release of the radioactive waste after disposal. Prior to the termination of the Yucca Mountain repository program, DOE was funding the development of "Transportation, Aging and Disposal" (TAD) SNF canisters. The goal of the TAD program was to allow spent fuel assemblies discharged from nuclear reactors to be placed into sealed canisters at the reactor sites and remain in the same sealed canisters through ultimate disposal in a deep geologic repository.[176] In addition, NRC has indicated that potential new fuels, such as fuels having different cladding, internal materials, different assembly designs, and different operating conditions, and fuels with higher burn-up than current limits may need further review to demonstrate that extended storage can be accomplished safely.[177]

As described above, some have expressed concern about the safety of wet pool storage of SNF. Utilities have continued to use wet pool storage largely because it is viewed as adequately safe, it is legally permitted, and shifting to dry cask storage for sufficiently cooled SNF would incur costs many would view as therefore unnecessary. Nuclear power plant operators must generally justify costs to stockholders and public utility commissions, as well as compete for grid dispatch based partly on costs. Accordingly, they seek to maximize the utilization of existing capacity for storage of SNF in wet storage pools before spending money on new, dry cask storage. Hence, the practice has generally been for utilities to move SNF to dry casks only when necessary after space in existing wet pools has been exhausted using approved dense packing procedures.[178] To do otherwise could be viewed by some as spending money unnecessarily before it is warranted or required. Two comparisons are possible in determining whether shifting to dry cask storage as soon as possible is warranted: (a) Does the current storage system "provide reasonable assurance of adequate protection?," and (b) Does dry cask storage provide sufficiently greater safety compared to wet storage pools to justify the cost and potential risks of transferring the SNF (i.e., comparing "wet to dry")?

Some have argued that the potential risks from dense packing of SNF in wet pools justifies a requirement to shift SNF to dry casks as soon as it has cooled.[179] Those arguments contend not merely that dry casks provide *better* protection, but that SNF storage pools fail to provide *adequate* protection, especially given the relatively large inventory of radioactive materials being stored in many SNF pools, compared to operating reactor cores. This concern was elevated after the disaster at the Fukushima Dai-ichi reactors that highlighted specific concerns about the potential for simultaneous failure sequences. Wet storage relies on active pumping and filtration systems that require electric power for operation. NRC regards SNF storage as generally posing little risk if effective precautions are taken. NRC's website describing the two SNF storage options concludes, "[t]he NRC believes spent fuel pools and dry casks both provide adequate protection of the public health and safety and the environment. Therefore there is no pressing safety or security reason to mandate earlier transfer of fuel from pool to cask."[180] Hence, compared to its mandate to provide "adequate protection," NRC has concluded that both technologies meet that goal.

Some have made a separate argument that dry casks provide safer SNF storage than wet pools. In its post-9/11 assessment of the safety and security of SNF storage, the National Academy of Sciences/National Research Council (NAS) Committee on the Safety and Security of Commercial Spent Nuclear Fuel Storage "judges that dry cask storage has several potential safety and security advantages over pool storage."[181] In the report's recommendation that has been cited frequently and interpreted diversely, NAS urged, "Depending on the outcome of plant-specific vulnerability analyses described in the committee's classified report, the [NRC] might determine that earlier movements of spent fuel from pools into cask storage would be prudent to reduce the potential consequences of terrorist attacks on pools at some commercial nuclear plants."

NRC's response to the NAS report focused on the question of whether wet pools and dry casks were *adequate*, rather than whether dry casks were *preferable*:

> [S]torage of spent fuel in both [wet storage pools] and in dry storage casks provides reasonable assurance that public health and safety, the environment, and the common defense and security will be adequately protected. The NRC will continue to evaluate the results of the ongoing plant-specific assessments and, based upon new information, would evaluate whether any change to its spent fuel storage policy is warranted.[182]
>
> The NRC believes spent fuel pools and dry casks both provide adequate protection of the public health and safety and the environment. Therefore there is no pressing safety or security reason to mandate earlier transfer of fuel from pool to cask.[183]

NRC has generally limited its statements on SNF storage options to whether they provide reasonable assurance of adequate protection, rather than comparing one technology with another. NRC Chairman Greg Jaczko was quoted, however, in a recent newspaper interview: "It's like the difference between buying one ticket in the Powerball lottery and 10 tickets."[184] Another NRC official went further, comparing the equivalency of dry cask storage to wet pool storage, saying,

"I think you have equal risk in both," and that federal policies consider both "equally safe, but they rely on different things to achieve that safety."[185]

The Nuclear Waste Technical Review Board concluded in 2010 that "the experience gained to date in the dry storage of spent fuel, demonstrates that used fuel can be safely stored in the short term and then transported for additional storage, processing, or repository disposal without concern."[186]

The merits of wet pool storage versus dry casks continue to be debated, with some arguing that the risks of continued storage in wet pools, at high-density configurations, requiring reliable power for cooling water, is too high a risk, while others argue that risks are relatively low, and the increased risk during the operation to move the SNF from the pools to dry casks exceeds the value of incremental safety improvement, and that the probabilities of a mishap with either technology are low and virtually indistinguishable.

Many have observed that the dry casks at the Fukushima Dai-ichi plant were unharmed despite the earthquake and subsequent tsunami, whereas the wet SNF storage pools that were located within the same containment structure as the reactors were damaged but not the wet storage pool outside the reactor building. The damage to the wet storage pools inside the reactor buildings occurred largely as a result of the hydrogen explosions and resulting structural collapse into the pools, exacerbated by the loss of power and subsequent lack of operating monitoring instrumentation in the pools.[187] Despite this extraordinary damage, the SNF in the pools appears to have been largely unharmed. Some nuclear critics argue that the magnitude of the radioactive material inventory in wet storage pools warrants protection from the risks of a possible release from any of a number of scenarios, including station blackout or an airplane strike.[188]

Some have argued that SNF should be transferred to dry storage after the necessary cooling-off period in wet pool storage. NRC has examined this issue and determined that transfer of SNF to dry storage is not necessary to provide reasonable assurance of adequate safety. Examining the technical merits for such a requirement is beyond the scope of this report. Most recently, NRC staff, in its Near-Term Task Force recommendations from the Fukushima Dai-ichi accident, did not recommend a requirement to transfer SNF from wet to dry storage.[189] NRC has issued orders requiring additional monitoring of SNF storage and safeguards against station blackout and lossof-power situations, which could add to the cost of maintaining wet pool storage.[190] These added costs, combined with the uncertainty about a permanent geologic repository, could increase the already accelerating rate of transfer from wet pools to dry storage. A shift of all SNF older than five years from wet pools to dry storage would more than triple the amount of SNF in dry cask storage.[191] If this shift were to occur, it would result in 85% of the total SNF stored in dry casks (instead of the current 25% of SNF in dry casks). To accommodate this number of transfers from wet storage pools, more than 2,700 new dry casks would be required.[192]

NRC currently provides specific technical requirements for both wet and dry cask storage, and points to this system as a reason there is no pressing safety or security reason to mandate earlier transfer of fuel from pool to cask (see "Hazards and Potential Risks Associated with SNF Storage" above).[193]

Because of the increasing use of dry cask storage, NRC's and industry's systems for implementing these requirements may require modification for a larger number of sites. A recent review by the NRC Office of Inspector General (OIG) evaluated NRC's process for inspecting the safety and security of dry cask storage. The OIG review did not identify any current technical risks with SNF storage using dry casks, but noted that NRC lacks an adequately centralized, consistent system for conducting dry cask inspection, including staffing and training. It found that NRC needed to address this issue because dry cask storage was the most likely method for additional storage capacity, and that all reactors are expected to require dry cask storage by 2025. As dry cask increasingly becomes the predominant method for SNF storage, rather than a supplement to wet pool storage, NRC would need to step up the level of staffing, training, and procedural discipline to ensure effective and consistent inspections.[194] The OIG findings did not identify any technical failures with dry casks, but rather recommended adjustments in NRC management to make oversight and inspections more systematic. NRC has begun implementing these improvements in the dry cask inspection program.[195]

NRC's recent *Near-Term Task Force Review of Insights from the Fukushima Dai-ichi Accident* did not include a mandate to switch from wet to dry storage among its recommendations for considerations by the commission. The Near-Term Task Force recommended, instead, "enhancing spent fuel pool makeup capability and instrumentation for the spent fuel pool" (Section 4.2.4) and a number of actions to address site blackout mitigation and preparedness.[196] The commission staff review of the task force recommendations subsequently found there was "no imminent risk from continued operation and licensing activities" and recommended action on several task force recommendations, including addressing site blackout issues (e.g., establish a "coping time of 72 hours for core and spent fuel pool cooling ..." under 10 C.F.R. 50.54(hh)(2)). The staff review recommended deferring action, however, on other task force recommendations related to SNF management, based on the need for additional review, in part to consider related regulatory actions, such as seismic qualification and instrumentation requirements.[197]

## Options for *Where* to Store SNF

The options for *where* to store SNF include alternatives ranging from continued decentralized storage at dozens of locations to some combination of consolidated or centralized storage facilities. Among the reasons for considering alternative storage locations are the costs incurred by multiple reactor locations, and concern that nuclear power plants were not originally located and designed to serve as indefinite SNF storage sites. In addition, some analysts have recommended long-term interim storage to improve safety and reduce pressure on establishing a repository in a forced technical and political environment.[198] The SNF storage option most often debated, and most recently recommended by the Blue Ribbon Commission (BRC), is whether to establish centralized storage locations, or consolidated storage for stranded SNF.

Any evaluation of options for where to store SNF begins with the current situation, which includes the location of existing SNF (see above section) and the statute governing SNF management, the Nuclear Waste Policy Act (NWPA). The current U.S. inventory of SNF is more than 67,000 MTU, and growing by approximately 2,000 MTU per year. Various studies have projected future inventories of SNF requiring storage and disposal. The results depend on assumptions about the rate at which SNF is generated, which depends on assumptions about reactor operations, license renewals, and new reactor construction. For example, a 1995 DOE study projected that SNF storage would reach approximately 86,000 MTU in about 2030 and level off unless a significant number of new reactors are ordered, constructed, and operated.[199] Future inventory levels, which depend on capacity factors at, and number of reactors in operation, and discharge rates for SNF, could have an impact on the need for storage and disposal capacity.

Secretary of Energy Spencer Abraham cited security concerns in the wake of the September 11, 2001, attacks as a basis for relocating high-level waste from at-reactor storage to Nevada surface storage.[200] Others have noted that at least some of the spent fuel and high-level waste will remain at the same temporary storage locations for as long as the facilities operate. Storage of discharged spent fuel will require at-reactor storage for at least one to five years to allow the SNF to cool sufficiently to allow transfer to a dry cask for storage or transport.

The BRC draft report noted that the NWPA allows for the construction of one consolidated interim storage facility with limited capacity, and only after a nuclear waste repository is licensed.[201] This statutory limitation reflects the perennial concern that establishment of a "temporary" or interim storage site could become a de facto permanent waste repository, and could reduce the perceived need for a permanent geological repository. BRC concluded that "[o]ne or more consolidated storage facilities will be required, independent of the schedule for opening a repository."[202] BRC explicitly recommended that the NWPA be "modified to allow for multiple storage facilities with adequate capacity to be sited, licensed and constructed when needed."[203] BRC's Transportation and Storage Subcommittee had earlier signaled its support for consolidated interim storage in its draft May 2011 report, which indicated that a central interim storage location could provide significant benefits, including reduced costs, improved safety, and increased overall confidence in nuclear power as a viable long-term energy source.[204]

The BRC recommendation is consistent with a major report earlier in the year by the Massachusetts Institute of Technology (MIT), which stated, "[t]he possibility of storage for a century, which is longer than the anticipated lifetimes of nuclear reactors, suggests that the U.S. should move toward centralized storage sites—starting with SNF from decommissioned reactor sites and in support of a long-term SNF management strategy."[205] Similarly, in testimony before BRC, a longtime nuclear analyst from Harvard University urged BRC to follow the advice of the bipartisan National Commission on Energy Policy, which recommended that Congress "[a]mend the Nuclear Waste Policy Act to make clear that interim storage and federal responsibility for waste disposal are sufficient to satisfy the Nuclear Regulatory Commission's waste [confidence] rule."[206] Except for state requirements, this need to address long-term SNF storage is not a near-term limitation on building new nuclear power plants.

Senators Lisa Murkowski and Mary Landrieu proposed legislation in June 2011 (S. 1320, the Nuclear Fuel Storage Improvement Act) intended to help establish up to two interim nuclear waste storage sites. The bill includes provisions to support DOE in paying local governments who offer to host interim spent fuel storage facilities using the Nuclear Waste Fund for these payments. It also directs DOE to arrange for the transportation of the SNF to the interim site.

The idea of consolidated interim storage has been debated for many years. The concerns about consolidated interim storage expressed by some include (1) potentially reducing pressure on decision makers to agree on a permanent repository because the problem will appear to have been "solved" and because of a resulting concern that storage sites would become de facto repositories; (2) protracted delays because of the difficulty in finding a willing host community for SNF storage, which could divert resources from other useful efforts (e.g., safety and security of existing SNF storage and repository siting); and (3) potentially higher net risks if the SNF were to require multiple moves to transport it to the interim site and again to the repository (depending on risk reduction from moving SNF from existing storage locations).

Many have found common ground on the need to consolidate storage of SNF that is located where a reactor or other nuclear facilities have been shut down (i.e., "stranded" SNF sites). If the SNF at these locations were consolidated to a new or existing storage site, it would have the effect of reducing the number of storage sites and reducing many fixed overhead costs (e.g., security and maintenance). It could also add a new storage site or sites to the map, depending on whether an existing storage site was used. There will be a temporary need for wet storage even after reactors at a site have ceased operation to allow time for the SNF to cool sufficiently (one to five years) to be transferred to dry cask storage.

Consolidating current inventories of SNF reduces, but cannot eliminate, the need for spent fuel at these sites. At sites where the reactor continues to operate, SNF will continue to be stored, at a minimum, to allow for cooling of newly generated SNF before it can be moved. Moving SNF from the 10 commercial "stranded" SNF sites[207] to a consolidation site would have the effect of reducing the number of sites requiring security and safeguards, thereby potentially reducing storage costs, or allowing more resources to be devoted to safety and security at the consolidated site, compared to multiple sites. There may be some incremental additional risk in transporting the SNF to a consolidated site. The risks and costs of continuing to store the SNF at the reactor site would have to be compared to the risks of transferring to another location. The United States has some significant experience with safely transporting SNF, largely through the DOE Foreign Research Reactor SNF return program and limited intra-utility transfers. In this program, DOE has, with no significant safety incidents, returned spent fuel from countries where the United States had previously shipped nuclear materials—often weapons-grade uranium—with the expectations that the material would be shipped back to the United States after being discharged from the reactor.

## Options for SNF Management and Regulatory Oversight

To consider *alternative management and regulatory oversight structures*, there are a range of possible options, including both government and quasi-governmental entities with varying authorities and resources. For decades, Congress has charged three federal agencies with the primary responsibilities[208] for high-level nuclear waste management, including SNF: DOE has been responsible for technical evaluations of the Yucca Mountain site, preparing license application documents and eventually operating the geologic repository; EPA has been responsible for developing standards; and NRC has been responsible for the licensing process, including the geologic repository and SNF storage in wet pools and dry casks. With the termination of the Yucca Mountain repository program at DOE, civilian SNF analysis has been transferred from DOE's former Office of Civilian Radioactive Waste Management to DOE's Office of Nuclear Energy.[209]

Various institutional structures and models have been proposed to manage nuclear waste in place of the current responsibility given to DOE, mandated by the NWPA. For example, Senator George Voinovich repeatedly introduced the U. S. Nuclear Fuel Management Corporation Establishment Act,[210] which would establish a government corporation with a bipartisan board of directors appointed by the President and confirmed by the Senate.

The Blue Ribbon Commission (BRC) concluded that "a new, single-purpose organization is needed to develop and implement a focused, integrated program for the transportation, storage, and disposal of nuclear waste in the United States," and that "moving responsibility ... outside DOE ... offers the best chance for future success."[211] Again, a similar recommendation had been included in the MIT report, which recommended "that a new quasi-government waste management organization be established to implement the nation's waste management program."[212] The BRC recommendation to change the institutional structure for managing SNF by creating a new federal corporation "dedicated solely to implementing the waste management program and empowered with the authority and resources to succeed"[213] was considered by some to be the "centerpiece of the BRC's recommendations,"[214] though they criticized it as falling short of their goal of a market-based private SNF management system.

Some have recommended the use of existing DOE nuclear weapons facilities for storage of SNF, coupled with spending on research for new reprocessing technologies.[215] The idea of establishing a consolidated storage facility for commercial SNF at a DOE facility is part of the larger debate about whether, how, and where to establish consolidated SNF storage sites. Proposals to use DOE sites (e.g., the Savannah River Site in South Carolina) for this purpose are often linked with using existing DOE facilities, including reprocessing "canyon" buildings. Although it is beyond the scope of this report, as Congress considers the BRC recommendations for alternative nuclear waste management structures, it may useful to consider prior federal government experience in operating and providing both support and regulatory oversight for disposal repository and reprocessing efforts.

Managing SNF storage and disposal through reprocessing, or chemical separations, is a perennial issue, distinct from the question of storage location, that has been promoted as an SNF solution for many years.[216] Others have argued that reprocessing has already been demonstrated to be not cost-effective, to produce significant environmental problems, and potentially to affect U.S. nuclear weapons nonproliferation policy. A variety of studies have concluded that reprocessing of SNF cannot be justified on the basis of waste management advantages or economics, but many support additional investment in research and development of more cost-effective reprocessing technologies.[217] A recent MIT study concluded, for example, "[f]or the next several decades, light water reactors using the once-through fuel cycle are the preferred option for the U.S." [218] These studies generally assume other options for spent fuel management, stockpiles and prices for uranium, and the readiness of the "fast" reactors that are viewed as a method to consume plutonium generated from reprocessing. Even the strongest advocates for reprocessing do not promote near-term deployment, and instead urge more government-funded research to pursue new reprocessing technologies, concluding "there is no benefit to reprocessing at this time."[219] BRC concluded that "no currently available or reasonably foreseeable reactor and fuel cycle technology developments—including advances in reprocessing and recycling technologies—have the potential to fundamentally alter the waste management challenge this nation confronts over at least the next several decades, if not longer."[220] Many analysts have raised concerns about the proliferation policy impacts associated with reprocessing, which gave rise to the U.S. halt in reprocessing by President Gerald Ford in the 1970s. Although this presidential ban was subsequently reversed, there are currently no plans for resumption of commercial reprocessing, largely because it has not been viewed as commercially viable.[221]

The conclusion that SNF reprocessing is not currently viable as a cost-effective alternative to storage and disposal does not necessarily mean that some viable SNF processing technology could not be developed in the future. Proponents of research on new SNF processing technologies assert that new SNF technologies would not necessarily generate the same wastes as in traditional aqueous "PUREX" reprocessing technologies, or pose a proliferation risk from separation of weapons-usable fissile materials. The Obama Administration has adopted and funded a policy consistent with the view of the chairman of the Senate Subcommittee on Clean Air and Nuclear Safety of the Environmental and Public Works Committee (Senator Thomas Carper), and the conclusions of an earlier MIT report that recommended research into new SNF processing technologies, and more fundamentally, a broader life-cycle effort to integrate fuel designs and reactors with long-term SNF disposition plans.

U.S. Spent Nuclear Fuel Storage 49

BRC recommended "increased federal funding be provided to the NRC to support ... ongoing work by the NRC to develop robust ... regulatory framework for advanced nuclear energy systems ... including NRC's ongoing review of the current waste classification system. Such a framework can help guide the design of new systems and lower barriers to commercial investment by increasing confidence that new systems can be successfully licensed."[222] The nuclear industry has indicated that NRC rulemaking is needed because "the cost of the technology cannot be determined until the regulatory framework is known."[223] In part to address industry concerns that the cost of the technology cannot be determined until the regulatory framework is known, NRC has initiated a rulemaking to establish a licensing process for SNF reprocessing. Some nuclear critics have questioned the need for rulemaking given the lack of proposals for reprocessing.[224] Many have noted that changing the nuclear fuel cycle system to accommodate reprocessing may require a change in law.

## ISSUES FOR CONGRESS

In the United States, SNF is largely stored at reactors where the SNF was generated, using dry casks as necessary when storage pool storage capacity is exhausted. Given current prospects for adequate repository capacity and existing and growing inventories of waste, long-term (60 years beyond reactor license) or extended (more than 150 years) of SNF storage is likely. While some have expressed concern about the safety of wet pool storage and preference for dry cask systems, others, including the NRC, believe that sufficient regulatory controls and economic incentives exist for ensuring safe storage of SNF. Regardless of actual or perceived risk, most growth in SNF storage has been with dry systems, and the portion of SNF storage using dry cask systems, compared to wet pools, is likely to continue to increase more rapidly. The trend is driven largely by the long-term cost advantages of dry cask systems in a storage regime that now has a longer time horizon.

The current SNF storage situation could change if Congress adopts the recommendations of the Blue Ribbon Commission on America's Nuclear Future. These recommendations address where and how SNF is stored, as well as the institutional structures responsible. Areas of significant uncertainty include the likelihood of success of siting a "temporary" centralized consolidated storage site, the establishment of a new agency or federal corporation to manage SNF, and the prospects for success in developing advanced nuclear fuel cycles that address the economic, environmental, and proliferation issues that have foiled past efforts. The Senate Energy and Water Development appropriations bill for FY2013 includes provisions to help fund efforts to adopt the BRC recommendations.[225] Also, the Senate Energy and Natural Resources Committee is reported to be developing legislation to address BRC recommendations.[226]

# APPENDIX. U.S. SPENT NUCLEAR FUEL STORAGE INVENTORIES

**Table A-1. Alphabetical List of 35 States Where Spent Nuclear Fuel is Stored As of December 31, 2011 (see Table 1 for data sorted by mass of SNF storage)**

| State (15 states have no stored SNF)a | Number of Facilities | Number of Sites | "Stranded" SNF Storage Sitesb | MTU | | | Assemblies | | | Table Notes |
|---|---|---|---|---|---|---|---|---|---|---|
| | | | | Wet Storage | Dry Cask | Total SNF | Wet Storage | Dry Cask | Total SNF | |
| Alabama | 5 | 2 | - | 2,647 | 540 | 3,187 | 10,978 | 2,180 | 13,158 | |
| Arizona | 3 | 1 | - | 1,052 | 903 | 1,955 | 2,490 | 2,136 | 4,626 | |
| Arkansas | 2 | 1 | - | 607 | 726 | 1,333 | 1,336 | 1,600 | 2,936 | |
| California | 7 | 4 | 2 | 2,017 | 916 | 2,933 | 4,750 | 2,486 | 7,236 | b & g |
| Colorado | 1 | 1 | 1 | 0 | 15 | 15 | - | 2,208 | 2,208 | b & f |
| Connecticut | 4 | 2 | 1 | 1,439 | 613 | 2,052 | 5,050 | 1,467 | 6,517 | b & g |
| Florida | 5 | 3 | - | 2,511 | 445 | 2,956 | 5,859 | 1,024 | 6,883 | f |
| Georgia | 4 | 2 | - | 2,018 | 592 | 2,610 | 7,366 | 3,264 | 10,630 | |
| Idaho | 1 | 1 | 1 | 50 | 81 | 131 | 144 | 177 | 321 | b |
| Illinois | 15 | 8 | 2 | 6,900 | 1,791 | 8,691 | 28,242 | 9,625 | 37,867 | b & c |
| Iowa | 1 | 1 | - | 259 | 217 | 476 | 1,452 | 1,220 | 2,672 | |
| Kansas | 1 | 1 | - | 646 | 0 | 646 | 1,434 | - | 1,434 | |
| Louisiana | 2 | 2 | - | 1,014 | 235 | 1,249 | 3,861 | 1,148 | 5,009 | |
| Maine | 1 | 1 | 1 | 0 | 542 | 542 | - | 1,438 | 1,438 | b |
| Maryland | 2 | 1 | - | 531 | 808 | 1,339 | 1,197 | 1,824 | 3,021 | f |
| Massachusetts | 2 | 2 | 1 | 542 | 122 | 664 | 3,082 | 533 | 3,615 | b |
| Michigan | 5 | 4 | 1 | 2,058 | 502 | 2,560 | 6,495 | 1,537 | 8,032 | b & f |
| Minnesota | 3 | 2 | - | 678 | 525 | 1,203 | 2,645 | 1,770 | 4,415 | g |
| Mississippi | 1 | 1 | - | 602 | 203 | 805 | 3,428 | 1,156 | 4,584 | |
| Missouri | 1 | 1 | - | 679 | 0 | 679 | 1,696 | - | 1,696 | |

| State (15 states have no stored SNF)a | Number of Facilities | Number of Sites | "Stranded" SNF Storage Sitesb | MTU | | | Assemblies | | | Table Notes |
|---|---|---|---|---|---|---|---|---|---|---|
| | | | | Wet Storage | Dry Cask | Total SNF | Wet Storage | Dry Cask | Total SNF | |
| Nebraska | 2 | 2 | - | 650 | 203 | 853 | 2,825 | 808 | 3,633 | g |
| New Hampshire | 1 | 1 | - | 455 | 93 | 548 | 944 | 192 | 1,136 | |
| New Jersey | 4 | 2 | - | 2,025 | 529 | 2,554 | 7,489 | 2,535 | 10,024 | h |
| New York | 7 | 4 | - | 3,082 | 495 | 3,577 | 12,466 | 1,820 | 14,286 | b & f |
| North Carolina | 5 | 3 | - | 3,018 | 544 | 3,562 | 10,612 | 1,480 | 12,092 | |
| Ohio | 2 | 2 | - | 1,083 | 34 | 1,117 | 4,542 | 72 | 4,614 | |
| Oregon | 1 | 1 | 1 | 0 | 345 | 345 | - | 801 | 801 | b |
| Pennsylvania | 9 | 5 | - | 4,606 | 1,459 | 6,065 | 20,898 | 8,424 | 29,322 | d |
| South Carolina | 8 | 5 | 1 | 2,236 | 1,808 | 4,044 | 5,001 | 3,896 | 8,897 | b & e |
| Tennessee | 3 | 2 | - | 1,095 | 470 | 1,565 | 2,386 | 1,024 | 3,410 | |
| Texas | 4 | 2 | - | 2,121 | 0 | 2,121 | 4,522 | - | 4,522 | |
| Vermont | 1 | 1 | - | 513 | 111 | 624 | 2,815 | 612 | 3,427 | |
| Virginia | 4 | 2 | - | 970 | 1,477 | 2,447 | 2,120 | 3,229 | 5,349 | f |
| Washington | 2 | 2 | 1 | 319 | 339 | 658 | 1,715 | 1,836 | 3,551 | b |
| Wisconsin | 4 | 3 | 1 | 915 | 419 | 1,334 | 2,603 | 1,088 | 3,691 | b & f |
| U.S. Total Commercial Site Storage | 119 | 74 | 10 | 46,733 | 15,859 | 62,592 | 165,583 | 56,493 | 222,076 | |
| U.S. Total DOE Site Storage | 4 | 4 | 4 | 2,605 | 2,243 | 4,848 | 6,860 | 8,117 | 14,977 | |

Source: The primary source for these data is the Nuclear Energy Institute (NEI) report, "2011 Used Fuel Data," prepared by Gutherman Technical Associates, January 14, 2012. Site-specific data on sites with no operating reactors ("Storage-only Sites") is derived largely from U.S. DOE, Report to Congress on the Demonstration of the Interim Storage of Spent Nuclear Fuel from Decommissioned Nuclear Power Reactor Sites, DOE/RW-0596, December 2008. Data for DOE sites were generally from Frank

Marcinowski, Overview of DOE's Spend Nuclear Fuel & High Level Waste; Presentation to the Blue Ribbon Commission on America's Nuclear Future, U.S. DOE, March 25, 2010.

[a] There are currently 15 states with no commercial SNF storage (there may be temporary and relatively small-scale storage of SNF from non-power generating research and academic reactors): Alaska, Delaware, Hawaii, Indiana, Kentucky, Montana, Nevada, New Mexico, North Dakota, Oklahoma, Rhode Island, South Dakota, Utah, West Virginia, Wyoming.

[b] "Stranded" is generally used to refer to SNF stored where the nuclear reactor that generated the SNF has ceased operating and been decommissioned, and the SNF remains at the site. In some cases the wet storage pools have been dismantled and the SNF is stored in dry casks. In the case of the Morris, IL, site, the "stranded" SNF was shipped from other reactor sites for a proposed reprocessing facility that never operated., and no reactor has ever operated at the site. The number of "stranded" SNF storage sites here does not include sites where SNF is stored at decommissioned reactor sites co-located with operating reactors (e.g., San Onofre (CA), Dresden 1 (IL), or Indian Point (NY)). The table does include DOE nuclear weapons facilities in Colorado, Idaho, South Carolina, and Washington that store commercial SNF—sites where DOE reactors largely ceased operations in the 1980s.

[c] Includes the Morris, IL, site, operated by GE-Hitachi, which never hosted an operating reactor or generated any SNF. The site was built to serve as an SNF reprocessing plant for which the SNF from other sites was shipped. The facility never operated, and the SNF has remained stored at the site. Many sources categorize the Morris site with dry cask storage sites, because they are all considered "Independent Spent Fuel Storage Installations" (ISFSI), although the Morris site differs from other ISFSIs because it uses wet pool storage and is categorized accordingly here.

[d] Excludes SNF and debris generated at the Three Mile Island-2 facility removed after the 1979 incident and shipped to DOE's Idaho National Engineering Laboratory (now know as the "Idaho National Laboratory").

[e] Includes 29 MTU of SNF, fragments, and nuclear materials stored at the Savannah River Site (SRS) near Aiken, which has been operated by DOE for nuclear weapons material production. In addition to material generated there, the SRS is used to store SNF shipped from commercial reactors (e.g., Carolinas-Virginia Tube Reactor) and from foreign and domestic research reactors using U.S.-origin fuel.

[f] Does not include SNF shipped to DOE federal facilities: 8 MTU from Florida, 1 MTU from Maryland, 12 MTU from Michigan, 16 MTU from New York, 22 MTU from Virginia, and 4 MTU from Wisconsin, as well as part of the SNF from the Fort St. Vrain site in Colorado, which was shipped to Idaho.

[g] Does not include SNF shipped to the Morris, IL, facility (see (b) above) from 4 states: 100 MTU from California, 34 MTU from Connecticut, 185 MTU from Minnesota, and 191 MTU from Nebraska. Includes 132 MTU shipped from other facilities in Illinois to the Morris facility.

[h] The adjacent Salem 1 and 2 and Hope Creek reactors share dry cask storage facilities and could be considered three facilities on a single contiguous site.

# End Notes

[1] "Spent Nuclear Fuel" (SNF) is sometimes referred to as "used" nuclear fuel. This difference in terminology often reflects a significant policy debate about whether SNF is a waste destined for disposal, or a resource. This report uses the term SNF to be consistent with other independent agency documentation. The broader term, "nuclear waste," which is sometimes used to refer to SNF, is generally not used here to avoid imprecision and confusion with other waste forms (e.g., low-level or transuranic waste and naturally occurring radioactive materials). Although they are quite different technically, spent nuclear fuel and liquid raffinate waste from reprocessing are both defined legally under the Nuclear Waste Policy Act (NWPA) as "high-level waste." See 42 U.S.C. 10101(2)(12) and 10 C.F.R. 60.2, 10 C.F.R. 63.2, and 40 C.F.R. 197.2.

[2] House Energy and Commerce Committee; Environment Subcommittee Hearing, "DOE's Role in Managing Civilian Radioactive Waste," June 1, 2011, http://energycommerce.house.gov/news/PRArticle.aspx?NewsID=8710; and House Energy and Commerce Subcommittee on Environment and the Economy, "Bipartisan Concern Over Administration's Haste to Terminate Permanent Nuclear Repository," press release, June 15, 2011.

[3] See CRS Report R40202, Nuclear Waste Disposal: Alternatives to Yucca Mountain, by Mark Holt; and CRS Report RL33461, Civilian Nuclear Waste Disposal, by Mark Holt, for discussion of related issues.

[4] See, e.g., National Research Council, Board on Radioactive Waste Management, Rethinking High-Level Radioactive Waste Disposal: A Position Statement of the Board on Radioactive Waste Management (1990), p. 2. National Academies/National Research Council/Board on Radioactive Waste Management, Disposition of High-Level Waste and Spent Nuclear Fuel: The Continuing Societal and Technical Challenges, National Academy Press, Washington, DC, 2001; Massachusetts Institute of Technology, The Future of the Nuclear Fuel Cycle: An Interdisciplinary MIT Study, MIT, Cambridge, MA, 2011; and Matthew Bunn, et al., Interim Storage of Spent Nuclear Fuel: A Safe, Flexible, and Cost-Effective Near Term Approach to Spent Nuclear Fuel Management, Harvard Press, Cambridge, MA, Harvard University Project on Managing the Atom and University of Tokyo Project on Nuclear Energy, 2001, http://belfercenter.ksg.harvard.edu/files/spentfuel.pdf.

[5] Blue Ribbon Commission on America's Nuclear Future, Report to the Secretary of Energy, January 2012, http://brc.gov/sites/default/files/documents/brc_finalreport_jan2012.pdf.

[6] Blue Ribbon Commission on America's Nuclear Future, U.S. Department of Energy, Advisory Committee Charter, filed with Congress March 1, 2010. See http://brc.gov/index.php?q=page/charter.

[7] Kraft, Steven P., "Used Nuclear Fuel Integrated Management," Nuclear Energy Institute, Presentation to National Association of Regulatory Utility Commissioners (NARUC), February 13, 2010; Redmond, Everett, "Nuclear Energy Institute Comments on the Disposal Subcommittee Draft Report to the Blue Ribbon Commission on America's Nuclear Future," July 1, 2011 (see http://brc.gov/sites/default/files/comments/attachments/ comments_of_nuclear_energy_institute_-_july_1_2011.pdf), and Kraft, Steven P., "Used Nuclear Fuel Integrated Management," Nuclear Energy Institute, Presentation to National Association of Regulatory Utility Commissioners (NARUC), February 13, 2010.

[8] Cochran, Thomas B. (Natural Resources Defense Council, Inc.), "Statement on the Fukushima Nuclear Disaster and its Implications for U.S. Nuclear Power Reactors," Joint Hearings of the Subcommittee on Clean Air and Nuclear Safety and the Committee on Environment and Public Works, U.S. Senate, April 12, 2011, and American Physical Society Nuclear Energy

Study Group, Consolidated Interim Storage of Commercial Spent Nuclear Fuel: A Technical and Programmatic Assessment, February 2007.

[9] Chairman Dianne Feinstein, Opening Statement, Hearing of the Senate Committee on Appropriations Subcommittee on Energy and Water Development, "U.S. Nuclear Power Safety in Light of Japan Disaster," March 30, 2011.

[10] Senator Dianne Feinstein, Letter to Gregory Jaczko, Chairman of U.S. Nuclear Regulatory Commission, April 8, 2011.

[11] Ibid.

[12] Senator Dianne Feinstein, Safety and Economics of Light Small Water Modular Reactors, Statement at Hearing on Light Small Water Modular Reactors, Senate Energy and Water Development Appropriations Subcommittee, June 14, 2011.

[13] U.S. House Energy and Water Development Appropriations Subcommittee, 112th Congress Report; 1st Session; Energy and Water Development Appropriations Bill Committee Report, 2012 (June 2011). Also see Committee on Science, Space, and Technology, Majority Staff Report; "Yucca Mountain: The Administration's Impact on U.S. Nuclear Waste Management Policy," June 2011.

[14] See generally, Walker, J. Samuel (NRC Historian), The Road To Yucca Mountain: The Development of Radioactive Waste Policy in the United States, University of California Press, 2009.

[15] See CRS Report R40202, Nuclear Waste Disposal: Alternatives to Yucca Mountain, by Mark Holt, and CRS Report RL34234, Managing the Nuclear Fuel Cycle: Policy Implications of Expanding Global Access to Nuclear Power, coordinated by Mary Beth Nikitin.

[16] International Atomic Energy Agency, Nuclear Power Reactors in the World, 2010 Edition (Reference Data Series No. 2), IAEA, Vienna, Austria. See http://www-pub.iaea.org/ MTCD/Publications/PDF/iaea-rds-2-30_web.pdf.

[17] Finland (with four nuclear reactors) is the only country where a commercial nuclear waste repository site has been selected with local government involvement and support, and characterization work has begun for a permanent high-level radioactive repository. Sweden has operated an interim central SNF pool storage facility since 1985 at Oskarshamn, and plans to develop a permanent geologic repository about 200 miles north at the Forsmark nuclear power plant near Östhammar, which has a right to veto the permit. France has a strongly supported nuclear power program, but has not yet selected a disposal site for the high-level waste and SNF (approximately 13,500 MT of SNF (as of 2007) and 2,229 cubic meters of vitrified high-level waste (as of 2007) were stored in France).

[18] "Transuranic" waste is defined as alpha-emitting radioactive waste contaminated with radionuclides heavier than uranium (atomic weight 93) at a concentration greater than 100 nCi/g (3.7 MBq/kg). The Waste Isolation Pilot Plant Land Withdrawal Act as amended by P.L. 104-201 (H.R. 3230, 104th Congress)."

[19] de Saillan, Charles, "Disposal of Spent Nuclear Fuel in the United States and Europe: A Persistent Environmental Problem," Harvard Environmental Law Review 34, 461 (2010).

[20] See generally McCutcheon, Chuck, Nuclear Reactions, The Politics of Opening a Radioactive Waste Disposal Site, University of New Mexico Press, 2002, ISBN-10: 0826322093.

[21] The opening and operation of WIPP is given by some as evidence that the "nuclear waste problem" can be solved. While recognizing the successful siting and permitting of WIPP, others note inherent uncertainty about long-term performance. Also, some have suggested that the WIPP site be used later for disposal of high-level waste and spent nuclear fuel, in addition to disposal of transuranic wastes for which it is currently dedicated. Such a change of mission at WIPP would require, at a minimum, amending Section 12 of the WIPP Land

# U.S. Spent Nuclear Fuel Storage

57

Withdrawal Act, which provides a "ban on high-level radioactive waste and spent nuclear fuel." P.L. 102-579, 106 Stat. 4777.

[22] An SNF storage "facility" is generally associated with a reactor (see Figure 2 and Figure 3), except in the case of the Morris, IL, facility, which was initially built not as a power reactor, but as part of an SNF reprocessing facility that never operated. In many cases there are more than one nuclear power reactor and SNF storage "facility" at a "site," which is a geographically contiguous area (see "Where Is Spent Nuclear Fuel Located Now in the United States?"). Because dry cask storage areas are typically independent of a reactor facility at a site, dry cask storage is generally quantified according to number of sites, not facilities.

[23] Feiveson, Harold, Zia Mian, M.V. Ramana and Frank von Hippel, International Panel on Fissile Materials, "Spent Fuel from Nuclear Power Reactors: An Overview of a New Study by the International Panel on Fissile Materials," September 2011. The study summarizes SNF management in the United States, Canada, France, Germany, Finland, Japan, South Korea, Russia, Sweden, and the United Kingdom. France provides the greatest support for reprocessing its SNF, but is listed as having the third highest amount of SNF (13,500 MT in 2007), behind the U.S. and Canada.

[24] See "Hazards and Potential Risks Associated with SNF Storage" below.

[25] Christopher A. Kouts, Principal Deputy Director, Office of Civilian Radioactive Waste Management, U.S. Department of Energy, "Yucca Mountain Program Status Update," July 22, 2008, p. 18.

[26] Metric tons of uranium (MTU) is a common unit of measurement for SNF, which allows for consistent measurement/estimation of fuel burnup. In some cases SNF is measured by metric tons of heavy metal or initial heavy metal (to distinguish the uranium fuel from the metal cladding), which have significant differences for evaluating reactor operations, but the distinction reflects a relatively minor difference for measurement purposes in this overview of SNF inventory.

[27] 42 U.S.C. §10134.(d).

[28] Most reprocessing in the United States was performed for separation of nuclear materials for nuclear weapons and other defense purposes. A relatively small amount of high-level radioactive waste was generated at, and remains stored at, the West Valley site in New York, after being vitrified in the 1990s. Most of the material reprocessed at the West Valley site was civilian SNF.

[29] U.S. Department of Energy Office of Civilian Radioactive Waste Management, Final Supplemental Environmental Impact Statement for a Geologic Repository for the Disposal of Spent Nuclear Fuel and High-Level Radioactive Waste at Yucca Mountain, Nye County, Nevada Summary, DOE/EIS-0250F-S1, June 2008.

[30] DOE, Office of Civilian Radioactive Waste Management, "The Report to the President and the Congress by the Secretary of Energy on the Need for a Second Repository," DOE/RW-0595; December 2008.

[31] Massachusetts Institute of Technology, The Future of the Nuclear Fuel Cycle: An Interdisciplinary MIT Study, MIT, Cambridge, MA, 2011, at page xi.

[32] The Atomic Energy Commission (AEC) was a predecessor to the U.S. Department of Energy (DOE). The Energy Research and Development Administration succeeded the AEC in 1975 and was replaced by DOE in 1978.

[33] Walker, J. Samuel (NRC Historian), The Road to Yucca Mountain: The Development of Radioactive Waste Policy in the United States, University of California Press, 2009; and

Atomic Energy Commission Press Release, Frank K. Pittman, "Management of Commercial High-Level Nuclear Radioactive Waste," July 25, 1972, AEC Press release.

[34] 42 U.S.C. 10242.

[35] Makhijani, Arjun, and Scott Saleska, High-Level Dollars, Low-Level Sense, Institute for Energy and Environmental Research, Apex Publishers, Takoma Park, MD 1991.

[36] Several states, including New York and Connecticut, have challenged NRC's "Waste Confidence" rule, which concluded that SNF "can be stored safely and without significant environmental impacts ..." (See United States Court of Appeals for the District of Columbia Circuit, State of New York, et al., Petitioners, Nos. 11-1045, 11-1051, -against- 11-1056, 11-1057 Nuclear Regulatory Commission and United States of America, Respondents); and Alvarez, Robert, Spent Nuclear Fuel Pools in the U.S.: Reducing the Deadly Risks of Storage, Institute for Policy Studies, May 2011; Zhang, Hui. Radiological Terrorism: Sabotage of Spent Fuel Pools, INESAP: International Network of Engineers and Scientists Against Proliferation no. 22 (December 2003): 75-78.

[37] A. G. Sulzberger and Matthew L. Wald, "Flooding Brings Worries Over Two Nuclear Plants," New York Times, June 20, 2011.

[38] U.S. Nuclear Regulatory Commission, "Consideration of Environmental Impacts of Temporary Storage of Spent Fuel After Cessation of Reactor Operations; Waste Confidence Decision Update; Final Rule," 75 Federal Register 81032, December 23, 2010; and National Academies/National Research Council, Committee on Climate Change and U.S. Transportation, Potential Impact of Climate Change on U.S. Transportation: Special Report 290, 2008.

[39] For more information on this liability issue, see CRS Report R40996, Contract Liability Arising from the Nuclear Waste Policy Act (NWPA) of 1982, by Todd Garvey.

[40] The claims have been paid from the "Judgment Fund," which is a permanent, indefinite appropriation for the payment of judgments against the United States managed by the U.S. Department of Justice (see 31 U.S.C. §1304), not from appropriations to DOE through the Energy and Water Development Subcommittees. The U.S. government cannot use the Nuclear Waste Fund to pay for any of the damages that the utilities incur as a result of DOE's delay, because on-site storage is not one of the uses of the NWF authorized by the NWPA (see Alabama Power Co. v. United States Department of Energy, 307 F.3d 1300 (11th Cir. 2002)).

[41] U.S. Government Accountability Office, GAO-11-731T, Nuclear Waste, Disposal Challenges and Lessons Learned from Yucca Mountain, June 1, 2011, http://www.gao.gov/ new.items /d11731t.pdf.

[42] The utilities' damages claims largely consist of the costs incurred to store SNF. The costs include capital costs to construct dry storage facilities or additional wet storage racks, costs to purchase and load casks and canisters, and costs of utility personnel necessary to design, license, and maintain these storage facilities.

[43] Court of Appeals for the Federal Circuit, in Maine Yankee Atomic Power Co. v. United States, 225 F.3d 1336, 1343 (Fed. Cir. 2000). and Pac. Gas & Elec. Co. v. United States, 536 F. 3d 1282, 1284, 1287 (Fed. Cir. 2008.)

[44] Michael F. Hertz, U.S. Department of Justice, Statement Before the Blue Ribbon Commission on America's Nuclear Future, February 2, 2011.

[45] American Physical Society Nuclear Energy Study Group, Consolidated Interim Storage of Commercial Spent Nuclear Fuel: A Technical and Programmatic Assessment, February 2007; Michael F. Hertz, Deputy Assistant Attorney General, Civil Division, Before the Committee on the Budget, U.S. House of Representatives, Budgeting for Nuclear Waste

Management, presented on July 16, 2009; and Hertz, Michael F., U.S. Department of Justice, Statement Before the Blue Ribbon Commission on America's Nuclear Future, February 2, 2011.

[46] Frank von Hippel, The Uncertain Future of Nuclear Energy, International Panel on Fissile Materials Report #9, September 2010.

[47] American Physical Society Nuclear Energy Study Group, Consolidated Interim Storage of Commercial Spent Nuclear Fuel: A Technical and Programmatic Assessment, February 2007.

[48] Luther J. Carter, Lake H. Barrett, and Kenneth C. Rogers, "Nuclear Waste Disposal Showdown at Yucca Mountain," Issues in Science and Technology (Lawrence Livermore National Laboratory), Fall 2010.

[49] Richard B. Stewart and Jane B. Stewart, Fuel Cycle to Nowhere: U.S. Law and Policy on Nuclear Waste (Nashville, TN: Vanderbilt University Press, 2011).

[50] Blue Ribbon Commission on America's Nuclear Future, Report to the Secretary of Energy, January 26, 2012, p. vi.

[51] Richard C. Moore, Enhancing the Role of State and Local Governments in America's Nuclear Future: An Idea Whose Time Has Come, prepared for the Blue Ribbon Commission on America's Nuclear Future, May 2011.

[52] Savage, Melissa, "The Other Nuclear Problem," State Legislatures Magazine (National Conference of State Legislatures), May 2011 (see http://www.ncsl.org/?tabid=22533). Also, see CRS Report R41984, State Authority to Regulate Nuclear Power: Federal Preemption Under the Atomic Energy Act (AEA), by Todd Garvey.

[53] See CRS Report R41984, State Authority to Regulate Nuclear Power: Federal Preemption Under the Atomic Energy Act (AEA), by Todd Garvey.

[54] See CRS Report RL32163, Radioactive Waste Streams: Waste Classification for Disposal, by Anthony Andrews (2006), and CRS Report RS22001, Spent Nuclear Fuel Storage Locations and Inventory, by Anthony Andrews (2004; available upon request).

[55] Uranium dioxide (UO2) is the most commonly used form of uranium used for fuel in commercial nuclear reactors, which is the focus of this report. Commercial reactor fuel uses uranium enriched to approximately 3%-5% U-235, with the balance of the uranium (approximately 95%-97%) U-238, which is "non-fissile," or incapable of sustaining a nuclear chain reaction. Other reactors for use in naval nuclear propulsion reactors, nuclear materials production reactors, and research applications may have different compositions in terms of uranium enrichment, overall structure, and cladding, but this report focuses on the typical fuel used in most commercial reactors. Some have proposed using thorium and/or metallic fuels (see, e.g., http://www.ltbridge.com) for "next generation" nuclear fuels.

[56] W. J. Bailey, A.B. Johnson, and D.E. Blahnik, et al., Surveillance of LWR Spent Fuel in Wet Storage, Battelle Pacific Northwest Laboratories, Electric Power Research Institute Report NP-3765, October 1984.

[57] The mass per assembly is based on the average total discharge mass (metric tons of heavy metal, or MTHM) reported by the Energy Information Administration (EIA) for BWR and PWR reactors from 1968 to 1998, which was a period for which complete and comparable records were available at Energy Information Administration, Form RW-859, "Nuclear Fuel Data" (1998). The quantity of spent fuel is measured in MTHM. Heavy metal refers to elements with atomic number greater than 89—in SNF, almost all heavy metal is uranium. The mass generally refers to the initial heavy metal (i.e., before irradiation), and is roughly equivalent to the MTU. After discharge, some (about 4%) of the uranium has been replaced by fission products.

[58] Light water reactors are distinguished from heavy water reactors, which use deuterium oxide (D20) instead of H20. Like H20, D2O is composed of two hydrogen atoms and an oxygen atom, except the hydrogen atoms each possess an extra neutron.

[59] The relatively large uranium-235 atom splits into smaller fission products, with different radiological and chemical characteristics, such as cesium-137 (half-life (t1/2) = 30 years), strontium-90 (t1/2 = 29 years), and technecium-99 (t1/2 = 211,000 years). It is one of the fundamentally unique and remarkable characteristics of nuclear fission that it is capable of creating wholly new elements from others in a process that had previously eluded humans for centuries, since the days of the ancient alchemists.

[60] Some have suggested that "reprocessing" (dissolving of solid spent fuel rods in acid and using organic solvents to separate long-lived plutonium and uranium isotopes from certain fission products) could provide an alternative to spent fuel disposal. In the United States, most reprocessing was conducted from the 1940s until 1992 to extract plutonium and other nuclear materials, primarily for nuclear weapons production. A complete examination of this technology is beyond the scope of this report and has been addressed in other CRS products. See, e.g., CRS Report RL34234, Managing the Nuclear Fuel Cycle: Policy Implications of Expanding Global Access to Nuclear Power, coordinated by Mary Beth Nikitin.

[61] The January 2011 NEI "Used Fuel" inventory report by Brian Gunterman estimated the average annual discharges of commercial SNF in the United States are slightly more than 2,000 MTHM (2,248 MTHM average annual rate for 2005- 2010, and 2,259 MTHM average annual discharge rate projected for 2005-2025), but 2,000 MTHM is the estimated annual rate most commonly used in other analyses and is used here to provide comparability.

[62] The production reactors were all built by the former Atomic Energy Commission and the original Manhattan Project in World War II at the Hanford Site in eastern Washington and the Savannah River Plant in South Carolina. Instead of generating heat to drive a turbine to generate electricity, these reactors generally used high-enriched uranium fuel to irradiate targets to transmute U-238 into weapons-grade Pu-239. Although commercial reactors also produce Pu-239, theft of commercial SNF is not directly a significant proliferation risk because the concentration of Pu-239 is relatively low (compared to other plutonium isotopes). Perhaps more significantly, extracting the most useful plutonium isotope for efficient weapons (i.e., Pu-239) requires a much more challenging step of reprocessing to separate the Pu-239 from the fission products, which otherwise renders the fissile material useless for power or weapons, thereby making commercial SNF a low proliferation risk. Another difference with DOE production reactor SNF, compared to commercial SNF, is that the fuel cladding is relatively thin-walled to facilitate reprocessing by reducing the material that has to be dissolved. One reactor ("N-reactor" at the Hanford site) was built to produce both weapons material and electricity. Although it was built in the early 1960s as a potential model for other "dual use" reactors, it was never replicated and shut down in the 1980s.

[63] Research reactor fuel is generally smaller than commercial fuel (i.e., about 4 feet compared to 12 to 15 feet long). In many cases, research reactors used high-enriched or "weapons grade" uranium and posed a greater proliferation threat than low-enriched commercial fuel. DOE has operated a Reduced Enrichment Research and Test Reactor Program to help these reactors convert to high-density, low-enriched uranium fuel.

# U.S. Spent Nuclear Fuel Storage 61

[64] The fuel used by the Naval Nuclear Propulsion Program has a significantly different and classified design from commercial, production, or research reactor fuel. Among other things, it uses high-enriched uranium and lasts considerably longer.

[65] Blue Ribbon Commission paper by DOE contractors, "Used Fuel Disposition: U.S. Radioactive Waste Inventory and Characteristics Related to Potential Future Nuclear Energy Systems," FCRD-USED-2011-000068, Rev. 18, May 2011.

[66] Dry cask storage is sometimes considered synonymous with "Independent Spent Fuel Storage Installation" (ISFSI) because all but one ISFSI site (GE Morris in Illinois) uses dry casks. This report uses both terms where appropriate. Also, this report generally refers to typical commercial dry cask storage. In fact, a wide range of dry storage configurations are used, including vertical and horizontal configurations (see Figure 4), as well as in-ground and indoor vaults, which are used disproportionately at DOE sites.

[67] 10 C.F.R. Part 50.

[68] 10 C.F.R. Part 72.

[69] Other, generally noncommercial, reactors types have been used for nuclear materials production by the Department of Energy and predecessor agencies for nuclear weapons material production, and for research purposes. In addition, commercial reactors in the former Soviet Union have used other designs, such as the graphite reactor (known as a "RBMK" reactor) used at the Chernobyl plant in the Ukraine.

[70] The Fort St. Vrain facility in Colorado (25 MT and 2,208 assemblies of SNF stored on-site, while the balance was shipped off-site to DOE's Idaho National Engineering Laboratory) was a high temperature gas-cooled reactor. The Santa Susanna Field Laboratory in California and the Fermi 1 reactor in Michigan were liquid sodium reactors. In addition to this historic variation in reactor designs in the United States, reactor and fuel designs in other countries vary significantly. Canada, for example, has used "CANDU" reactors that use heavy water, and the British have used "Magnox" reactors, with relatively unstable fuel intended for relatively quick reprocessing.

[71] The NRC Near-Term Task Force report in July 2011 put the number of assemblies in this basin at 6,291, while the INPO report from November 2011 indicated 6,375 assemblies were located in this separate storage pool. See U.S. NRC, Recommendations for Enhancing Reactor Safety in the 21st Century: The Near-Term Task Force Review of Insights from the Fukushima Dai-ichi Accident, July 12, 2011 (http://pbadupws.nrc.gov/docs/ML1118/ ML111861807.pdf); and Institute for Nuclear Power Operations, Special Report on the Nuclear Accident at the Fukushima Daiichi Nuclear Power Station, INPO 11-005, Rev. 0, November 2011, at page 35 (http://www.nei.org/ resourcesandstats/documentlibrary/ safetyandsecurity/reports/special-report-on-the-nuclear-accident-at-the-fukushimadaiichi-nuclear-power-station).

[72] In contrast to the practice of storing SNF in dry casks on outdoor storage pads in the United States, Germany requires dry casks to be stored in hardened storage buildings. See Ulrich Alter (Federal Ministry for the Environment and Nuclear Safety), Management of Radioactive Waste and Spent Fuel in Germany, presentation to the IAEA Conference, May 31, 2010.

[73] IAEA, "Operation and Maintenance of Spent Fuel Storage and Transportation Casks/ Containers," IAEA-TECDOC1532, January 2007.

[74] See http://www.iaea.org/inis/collection/NCLCollectionStore/_Public/37/088/37088622.pdf.

[75] See "Safety & Security of Commercial Spent Nuclear Fuel Storage: Public Report," National Academies Press, Washington, DC (2006).

[76] See http://www.nrc.gov/waste/spent-fuel-storage/designs.html.

[77] For example, the common SNF storage pool at the Fukushima site, in which about 60% of the SNF at the site was stored, was estimated to be capable of losing power for at least 30 days before it caused a concern. See Institute for Nuclear Power Operations, Special Report on the Nuclear Accident at the Fukushima Daiichi Nuclear Power Station, INPO 11-005, Rev. 0, November 2011, at page 35 (http://www.nei.org/resourcesandstats/documentlibrary/safetyandsecurity/reports/special-report-on-the-nuclear-accident-at-the-fukushima-daiichi-nuclear-power-station.)

[78] See generally http://www.nrc.gov/info-finder/reactor/na1.html.

[79] The United States signed the treaty on September 29, 1997, and ratified it on April 15, 2003. See http://www.iaea.org/Publications/Documents/Conventions/jointconv_status.pdf.

[80] International Atomic Energy Agency, Joint Convention on the Safety of Spent Fuel Management and on the Safety of Radioactive Waste Management, INFCIRC/546, December 24, 1997. See http://www.iaea.org/Publications/ Documents/Infcircs/1997/infcirc546.pdf.

[81] 10 C.F.R. 72.2 and see U.S. NRC, "NRC Standard Review Plan for Dry Cask Storage Systems" (NUREG-1536).

[82] Two types of dry storage are used at the site: 27 vertical Areva/Transnuclear (TN)-32 metal casks and 26 TN NUHOMS HD-32PTH horizontal storage modules. See http://www.nrc.gov/about-nrc/emerg-preparedness/virginiaquake-info/north-anna-isfsi-summary.pdf; and Kenneth D. Kok, Nuclear Engineering Handbook, CRC press, 2009.

[83] Gutherman Technical Services (hereinafter referred to as the NEI Report), "2010 Used Fuel Data," Report to Nuclear Energy Institute (Marcus Nichol), January 18, 2010.

[84] ACI Nuclear Solutions (Brian Gutherman), Report to Nuclear Energy Institute (Everett Redmond), 2009 Used Fuel Data, January 22, 2010.

[85] The term "sites" is used here to refer to geographically distinct locations. This accounting excludes the four sites where DOE is responsible for SNF storage in Colorado, Idaho, South Carolina, and Washington. Also, Hope Creek and Salem 1&2 (New Jersey) are counted as a single site because they are adjacent, located on the same artificial island in the Delaware River, share the same owner (PSEG Nuclear and Exelon), and share a dry cask storage pad using the same dry cask technology (Holtec Hi-Storm). An SNF storage "facility" is generally associated with a reactor that produced the SNF (see Figure 2 and Figure 3), except in the case of the Morris, IL, facility, which was initially built not as a power reactor, but as part of an SNF reprocessing facility that never operated. In many cases there are more than one nuclear power reactor and SNF storage facility at a "site," which is a geographically contiguous area. Because dry cask storage areas are typically independent of a reactor facility at a site, and have a licensing process separate from the reactor licensing, dry cask storage is generally quantified according to number of sites, not facilities.

[86] U.S. NRC, Draft Report for Comment: Background and Preliminary Assumptions For an Environmental Impact Statement—Long-Term Waste Confidence Update, December 2011, pp. 6-8 and 14.

[87] Excludes commercial SNF stored at DOE facilities.

[88] Includes the Hitachi-GE site in Morris, IL, but excludes the four DOE-operated sites.

[89] This number excludes three sites (Savannah River Site (SC) and Hanford (WA) and Fort St. Vrain (CO)) where DOE-owned SNF only is stored in dry storage facilities. The Idaho National Laboratory (INL) DOE site is included because some of the SNF stored there is from the Three Mile Island reactor, and the dry storage facility at INL for the TMI SNF debris is NRC-licensed.

# U.S. Spent Nuclear Fuel Storage 63

[90] Although the NEI data indicate it lacks a dry cask storage facility, the Salem (NJ) site is not included in this accounting because it is adjacent to the Hope Creek site and shares the dry cask storage facility there. Also, the Morris Site (IL) is included here because it is a wet storage site. In some databases, Morris is mistaken for dry casks because it is classified as an "Independent Spent Fuel Storage Installation (ISFSI)," which involves dry cask storage at all of the other ISFSI sites.

[91] Included nine former commercial reactor sites and the GE-Hitachi site in Morris, IL.

[92] The four DOE sites with long-term SNF storage are the Hanford Site (WA), Savannah River Site (SC), the Idaho National Laboratory (ID), and the Fort St. Vrain Site (CO). In addition, DOE has indicated that relatively small amounts of SNF (less than 50 kg at each) are stored at other sites including the Argonne National Laboratory (IL) and the Babcock and Wilcox facility in Lynchburg, VA. Also, several university research reactors temporarily store SNF. A total of about 3 MTU of SNF is stored at these various sites for which DOE provides support, away form the four main sites.

[93] One exception is the GE-Hitachi site in Morris, IL. This storage site never had a reactor site, but was slated to be a reprocessing site, and never operated.

[94] In addition to these former commercial nuclear power reactor sites, SNF is stored at the Morris, IL site. The Morris site may be considered "stranded" although it never hosted an operating nuclear reactor, but rather was intended for an SNF reprocessing plant (that never operated), for which the SNF was shipped.

[95] The term "stranded" is used to refer to situations where SNF is stored but the associated reactor is no longer operating and generating revenue to help pay for storage costs, despite the presence of associated reactors at the same site.

[96] In addition to the three former nuclear weapons material production facilities, DOE is also responsible for the storage of SNF located at the former Fort St. Vrain reactor site in Colorado, which now hosts a commercial natural gas power plant.

[97] Blue Ribbon Commission on America's Nuclear Future, Report to the Secretary of Energy, January 2012. http://brc.gov/sites/default/files/documents/brc_finalreport_jan2012.pdf. See page 35.

[98] Approximately two-thirds of the nuclear power reactors built in the United States have been PWRs, while about one-third have been BWRs. A disproportionate number of the reactors built in the west (72%) were PWRs, whereas a somewhat larger percentage of the reactors built in the East were BWRs (36%). Approximately 80% of the total number of reactors built in the United States were built in the East, compared to about 25% in the West. Of the 40 BWRs built in the United States, more than 80% were in the East. Of the 76 PWRs built, however, about 75% were built in the eastern half of the United States.

[99] Excludes two 132-liter drums of SNF stored at the West Valley site near Buffalo, NY. See Margaret Loop and Laurene Rowell, "Getting West Valley Demonstration Project Waste in the Right Path to Disposal," Waste Management 2011, February 27, 2011.

[100] Atomic Energy Commission and the Energy Research and Development Administration.

[101] The Idaho National Laboratory, known previously as the Idaho National Engineering Laboratory.

[102] U.S. DOE, Spent Fuel Storage Requirements 1994-2042, DOE/RW-0431-Rev.1, June 1995; and United States Court of Appeals for the District of Columbia Circuit, State of New York, et al.; Petitioners, Nos. 11-1045, 11-1051, -against- 11-1056, 11-1057 Nuclear Regulatory Commission, and United States of America, Respondents.

[103] U.S. DOE, The 1996 Baseline Environmental Management Report, USDOE/EM-0290, June 1996.

[104] Frank Marcinowski, Overview of DOE's Spent Nuclear Fuel & High Level Waste; Presentation to the Blue Ribbon Commission on America's Nuclear Future, U.S. DOE, March 25, 2010.

[105] These reactors, and the nuclear driver fuel and target assemblies in them, were designed for production of nuclear materials including weapons grade (Pu-239) and other forms of plutonium (e.g., Pu-238 for durable power sources), tritium, and neptunium.

[106] U.S. DOE Idaho Office, Kathleen Hain, Idaho Site Spent Nuclear Fuel; Management, Presentation to the Nuclear Waste Technical Review Board, June 2010; U.S. DOE Inspector General Office of Audit Services, Management of Spent Nuclear Fuel at the Savannah River Site (DOE/IG-0727), May 2006; U.S. DOE, Savannah River Site, Spent Nuclear Fuel Management Environmental Impact Statement (DOE/EIS-0279), March 2000; and U.S. DOE (Dawn Gillas), SRS Used Nuclear Fuel Management: A Presentation to the Citizens Advisory Board, July 2011.

[107] Thomas B. Cochran. Robert S. Norris, W. Arkin, and M. Hoenig (Natural Resources Defense Council), Nuclear Weapons Databook, Vol. II, U.S. Nuclear Warhead Production, Ballinger Publishing Company, Cambridge, MA, 1987. See http://docs.nrdc.org/nuclear /files/nuc_87010103a_65c.pdf.

[108] Largely core debris and damaged SNF from the Three Mile Island reactor. This was generally an exceptional situation, and most commercial reactor SNF is not sent to DOE facilities.

[109] Douglas Morrell, DOE Research Reactor Infrastructure Program: 2011 Status Report, September 14, 2011. See https://secure.inl.gov/TRTR2011/ Presentations/Morrell_ TRTR2011.pdf.

[110] For example, core debris resulting from the 1979 incident at the Three Mile Island-2 reactor in Pennsylvania, and commercial power demonstration projects at Shippingport and Peach Bottom in Pennsylvania, the Carolinas-Virginia Tube Reactor in South Carolina, Fort St. Vrain reactor in Colorado, and Sodium Reactor Experiment at the Santa Susanna Field Laboratory in California.

[111] Defense Nuclear Facilities Safety Board, "Review of the Hanford Spent Nuclear Fuel Project Defense Nuclear Facilities Safety Board," DNFSB/TECH-17, Technical Report, October 1997; Defense Nuclear Facilities Safety Board, "Technical Report: Stabilization of Deteriorating Mark 16 and Mark 22 Aluminum-Alloy Spent Nuclear Fuel at the Savannah Rover Site," DNFSB/TEC-7, June 1, 1995; and Defense Nuclear Facilities Safety Board, "Recommendation to the Secretary of Energy 94-1; Improved Schedule for Remediation of Nuclear Materials in the Defense Nuclear Facilities Complex," May 26, 1994.

[112] See CRS Report RS21988, Radioactive Tank Waste from the Past Production of Nuclear Weapons: Background and Issues for Congress, by David M. Bearden and Anthony Andrews.

[113] FY 2001 Defense Authorization Act, P.L. 106-398, Section 3137.

[114] Defense Nuclear Facilities Safety Board, "Recommendation to the Secretary of Energy 94-1."

[115] DOE National Nuclear Security Administration, "Press Release: NNSA Announces H-Canyon to Support Plutonium Disposition at the Savannah River Site," October 31, 2011 (see http://nnsa.energy.gov/mediaroom/pressreleases/ hcanyon); and George Lobsenz, "NNSA to Process Plutonium for MOX at H Canyon," Energy Daily, November 2, 2011.

[116] See CRS Report RS21988, Radioactive Tank Waste from the Past Production of Nuclear Weapons: Background and Issues for Congress, by David M. Bearden and Anthony Andrews.

[117] See National Academy of Sciences/National Research Council, "Risk Assessment in the Federal Government: Managing the Process," Committee on the Institutional Means for

# U.S. Spent Nuclear Fuel Storage

Assessment of Risks to Public Health, National Academy Press, Washington, DC, 1983; and Klaassen CD, Amdur MO, Doull J., Casarett & Doull's Toxicology: The Basic Science of Poisons, Macmillan Publishing (3rd Ed, 1986).

[118] Alvarez, Robert, Spent Nuclear Fuel Pools in the U.S.: Reducing the Deadly Risks of Storage, Institute for Policy Studies, May 2011.

[119] Wald, Matthew L., "Edward McGaffigan, 58, Atomic Official," New York Times, September 4, 2007. The term "adequate protection" is a term of art under Section 182 of the Atomic Energy Act of 1954. 42 U.S.C. 2232.

[120] Lisbeth Gronlund, David Lochbaum, and Edwin Lyman, Nuclear power in a warming world; Assessing the Risks, Addressing the Challenges, Union of Concerned Scientists, December 2007; Lochbaum, David, Nuclear Waste Disposal Crisis, Penn Well Books, Tulsa OK, 1996; Newsday Editorial, "Time to act on perilous piles; Memo to Congress: Do something now about our spent nuclear fuel," Newsday, April 22, 2011.

[121] After several years in pool storage, the thermal heat load from SNF drops significantly, and less heat removal from the water may be required. Nonetheless, the water continues to serve an important radiation shielding function.

[122] Jaczko, Honorable Greg, NRC Chairman, Statement before the Joint Hearing of the Subcommittee on Energy and Power and the Subcommittee on Environment and the Economy of the House Energy and Commerce Committee, "The FY2012 Department of Energy and Nuclear Regulatory Commission Budgets," March 16, 2011, http:// energy commerce.house.gov/hearings/hearingdetail.aspx?NewsID=8329.

[123] Institute for Nuclear Power Operations, Special Report on the Nuclear Accident at the Fukushima Daiichi Nuclear Power Station, INPO 11-005, Rev. 0, November 2011, at page 35; http://www.nei.org/resourcesandstats/ documentlibrary/safetyandsecurity/ reports/ special-report-on-the-nuclear-accident-at-the-fukushima-daiichi-nuclearpower-station.

[124] U.S. Nuclear Regulatory Commission, Information Notice 2011-05, Tohoku-Taiheiyou-Oki Earthquake Effects on Japanese Nuclear Power Plants, March 18, 2011.

[125] U.S. NRC, "Order for Interim Safeguards and Security Compensatory Measures," EA-02-026, February 25, 2002 (designated Safeguards Information, protected as "Official Use Only-Security Related Information").

[126] Ibid.

[127] Massachusetts Institute of Technology, The Future of the Nuclear Fuel Cycle: An Interdisciplinary MIT Study, MIT, Cambridge, MA, 2011.

[128] U.S. Government Accountability Office, DOE Nuclear Waste: Better Information Needed on Waste Storage at DOE Sites as a Result of Yucca Mountain Shutdown, GAO-11-230, March 2011.

[129] Lipa, Christine A., U.S. NRC, NRC Inspection Report No. 072-00001/11-01(Dnms)—General Electric-Hitachi Morris, Letter to Mark Varno, GE-Hitachi Global Laser Enrichment, LLC, May 13, 2011.

[130] Institute for Nuclear Power Operations, "Special Report on the Nuclear Accident at the Fukushima Daiichi Nuclear Power Station," INPO 11-005, Rev. 0, November 2011, at page 35 (http://www.nei.org/resourcesandstats/ documentlibrary/safetyandsecurity/ reports /special-report-on-the-nuclear-accident-at-the-fukushima-daiichi-nuclearpower-station.)

[131] For example, see U.S. NRC, Technical Study of Spent Fuel Pool Accident Risk at Decommissioning Nuclear Power Plants, NUREG-1738, 2001.

[132] National Academy of Sciences, National Research Council, Safety and Security of Commercial Spent Nuclear Fuel Storage: Public Report, National Academies Press 2006 (based on classified July 2004 report to Congress), at page 44.

[133] Letter from Nils Diaz, Commissioner, U.S. NRC to Senator Pete Domenici, with accompanying report: U.S. Nuclear Regulatory Commission, Report to Congress on the National Academy of Sciences Study on the Safety and Security of Commercial Spent Nuclear Fuel Storage, March 14, 2005.

[134] Robert Alvarez, Spent Nuclear Fuel Pools in the U.S.: Reducing the Deadly Risks of Storage, Institute for Policy Studies, May 2011.

[135] David Lochbaum, Nuclear Waste Disposal Crisis (Penn Well Books, Tulsa OK, 1996); Newsday Editorial, "Time to act on perilous piles; Memo to Congress: Do something now about our spent nuclear fuel," Newsday, April 22, 2011.

[136] National Academy of Sciences, National Research Council, Safety and Security of Commercial Spent Nuclear fuel Storage: Public Report (2006), E725-846.

[137] Ibid. at 36.

[138] Section 6 para. 2 no. 4 (German) Atomic Energy Act.

[139] Ulrich Alter (Federal Ministry for the Environment and Nuclear Safety), Management of Radioactive Waste and Spent Fuel in Germany, presentation to the IAEA Conference, May 31, 2010.

[140] Jussofie, A., R. Graf, W. Filbert, German Approach to Spent Fuel Management, "IAEA-CN-184/30, 2010.z; and Harold Feiveson, Zia Mian, M. V. Ramana, and Frank von Hippel, "Managing Nuclear Spent Fuel: Policy Lessons from a 10-country Study," Bulletin of the Atomic Scientists, June 27, 2011.

[141] "Cladding" is the zirconium alloy metal tube in which the ceramic uranium oxide fuel pellets are encased to make a fuel rod.

[142] "Zirconium has a high affinity for hydrogen. Absorption of hydrogen leads to hydrogen embrittlement, which can lead to failure of the zirconium tubing used as cladding for nuclear fuel. In addition, zirconium also reacts with oxygen, which can lead to corrosion." From Kenneth D. Kok, Nuclear Engineering Handbook (CRC press, 2009), p. 287. Also, see W. L. Hurt, K. M. Moore, E. L. Shaber, and R. E. Mizia, Extended Storage for Research and Test Reactor Spent Fuel for 2006 and Beyond, IAEA Conference on Extended Storage of Research Reactor Fuel, October 14, 1999– October 17, 1999, INEEL/CON-99-01083.

[143] The design of the fuel elements for U.S. commercial power reactors was fundamentally different from nuclear materials production reactors (e.g., Hanford and Savannah River sites plutonium production reactors), which involve high-enriched driver fuel surrounded by "target" billets with relatively thin-walled cladding to facilitate reprocessing to extract the plutonium or other nuclear materials. See U.S. DOE, Linking Legacies, DOE/EM-319, January 1997. Some have argued that British Magnox reactor fuel was also not designed for long-term storage. Others have argued that U.S. commercial fuels were also not designed for long-term storage and that the difference is more one of degree than of kind.

[144] Massachusetts Institute of Technology, The Future of the Nuclear Fuel Cycle: An Interdisciplinary MIT Study, MIT Press, Cambridge, MA, 2011.

[145] 49 Federal Register 34658 (August 31, 1984).

[146] Blue Ribbon Commission on America's Nuclear Future, Report to the Secretary of Energy, January 26, 2012, p. 89.

[147] EPRI, Used Fuel and High-Level Radioactive Waste Extended Storage Collaboration Program: November 2009 Workshop Proceedings, March 10, 2010; and Kessler, J., EPRI, Technical Bases for Extended Dry Storage of Spent Nuclear Fuel, 1003416 Final Report, December 2002.

[148] Personal e-mail from NRC staff (via Jenny Weil) to CRS, July 26, 2011.

# U.S. Spent Nuclear Fuel Storage

67

[149] U.S. NRC, List of Historical Leaks and Spills at U.S. Commercial Nuclear Power Plants, Rev. 7, June 14, 2011. See http://pbadupws.nrc.gov/docs/ML1012/ML101270439.pdf.

[150] In the case of widely reported tritium contamination of groundwater at the Vermont Yankee plant, the tritium release was determined to be from underground piping conduit leaks, not SNF storage.

[151] SNF storage pools were found to be one of a number of sources of leaks, including underground piping, valves on effluent discharge lines, leakage of components, and operator actions, which have caused several inadvertent releases. U.S. NRC, Liquid Radioactive Release—Lessons Learned Task Force Final Report, September 1, 2006, http://pbadupws.nrc.gov/docs/ML0626/ML062650312.pdf; see page 3.

[152] U.S. Nuclear Regulatory Commission, NRC Groundwater Task Force Final Report, June 2010. See http://pbadupws.nrc.gov/docs/ML1016/ML101680435.pdf.

[153] U.S General Accounting Office, Information on the Tritium Leaks and the Contractor Dismissal at the Brookhaven National Laboratory, GAO-RCED-98-26 (1998), and U.S. DOE, The 1996 Baseline Environmental Management Report, DOE/EM-0290, 1996.

[154] Nuclear Energy Institute Tritium Monitoring web page, "Fact Sheet: Industry Closely Monitors, Controls Tritium at Nuclear Power Plants," June 2009, http://www.nei.org/resourcesandstats/documentlibrary/safetyandsecurity/factsheet/ industrycloselymonitors controlstritium/.

[155] U.S. NRC, Regulatory Guide 4.1, "Radiological Environmental Monitoring for Nuclear Power Plants, Rev. 2" (June 2009, Rev. 2), http://pbadupws.nrc.gov/ docs/ML0913/ML091310141.pdf, and Regulatory Guide 1.21, "Measuring, Evaluating, and Reporting Radioactive Material In Liquid And Gaseous Effluents and Solid Waste, Rev. 2" (June 2009), http://pbadupws.nrc.gov/docs/ML0911/ML091170109.pdf.

[156] NRDC v. NRC, 582 F.2d 166 (2d Cir. 1978) and State of Minnesota v. NRC, 602 F.2d 412 (D.C. Cir. 1979).

[157] First waste confidence rule (1984): 49 Federal Register 34658 (August 31, 1984). NRC also amended 10 C.F.R. 51.23(a) to reflect this determination. Second waste confidence rule (1990): 55 Federal Register 38474 (September 18, 1990). Third waste confidence rule (2010): U.S. Nuclear Regulatory Commission, "Consideration of Environmental Impacts of Temporary Storage of Spent Fuel After Cessation of Reactor Operations; Waste Confidence Decision Update; Final Rule," 75 Federal Register 81032, December 23, 2010. NRC also amended 10 C.F.R. 51.23(a) to reflect this determination.

[158] NRC's first (1984) waste confidence rule predicted that "one or more repositories for commercial high-level radioactive waste and spent fuel will be available by the years 2007-2008." This determination was based in part on DOE's projection that the first repository would be available in 1998 and the second repository would open by 2004 (See DOE, Attachment to May 21, 1984 Letter from J. William Bennett, DOE to Robert W. Browning, NRC, "Draft Mission Plan," April 1984. NRC's second (1990) waste confidence rule predicted "at least one mined geologic repository will be available within the first quarter of the twenty-first century...."

[159] U.S. NRC, Tison Campbell, NRC Attorney, "A Level of Confidence," February 17, 2011, http://public-blog.nrcgateway.gov/2011/02/17/a-level-of-confidence/.

[160] In many cases this results in 120 years of expected safe storage if this 60 years (after the reactor operating period) is added to 60 years of storage while the reactor is operating.

[161] 10 C.F.R. 51.23 (a), and U.S. Nuclear Regulatory Commission, "Consideration of Environmental Impacts of Temporary Storage of Spent Fuel After Cessation of Reactor

Operations; Waste Confidence Decision Update; Final Rule," 75 Federal Register 81032, December 23, 2010.

[162] Kraft, Steven P., Nuclear Energy Institute, Used Nuclear Fuel Integrated Management, presentation to National Association of Regulatory Utility Commissioners (NARUC), February 13, 2010.

[163] 10 C.F.R. 51.23(b).

[164] United States Court of Appeals for the District of Columbia Circuit, State of New York, et al.; Petitioners, Nos. 11- 1045,11-1051, -against- 11-1056, 11-1057 Nuclear Regulatory Commission, and United States of America, Respondents.

[165] Christine Pineda, U.S. NRC, Waste Confidence Update for Long-term Storage, October 4, 2011. See http://pbadupws.nrc.gov/docs/ML1127/ML11276A152.pdf.

[166] U.S. NRC, Catherine Haney, Director of NRC Office of Nuclear Material Safety and Safeguards, Plan for the Longterm Update to the Waste Confidence Rule and Integration with the Extended Storage and Transportation Initiative, SECY-11-0029, February 28, 2011.

[167] Statement By Gregory B. Jaczko, Chairman, United States Nuclear Regulatory Commission to the House Committee on Energy and Commerce, Subcommittees on Energy and Power, and Environment and the Economy, March 16, 2011.

[168] Statement by Gregory B. Jaczko, Chairman United States Nuclear Regulatory Commission to the House Committee on Energy and Commerce Subcommittees on Energy and Power, and Environment and the Economy, May 4, 2011.

[169] U.S. NRC, Draft Report for Comment: Background and Preliminary Assumptions For an Environmental Impact Statement—Long-Term Waste Confidence Update, December 2011, pp. 6-8 and 14; U.S. NRC, Policy Issue Information Memorandum For The Commissioners from Catherine Haney, Plan for the Long-Term Update to the Waste Confidence Rule and Integration with the Extended Storage And Transportation Initiative, SECY-11-0029, February 28, 2011; U.S. NRC Memorandum to Commissioners from R. W. Borchardt, Project Plan for the Regulatory Program Review to Support Extended Storage and Transportation of Spent Nuclear Fuel, COMSECY-10-0007, June 15, 2010; subsequently approved by U.S. NRC Memorandum to R. W. Borchardt from Annette L. Vietti-Cook, Subject: Staff Requirements, Project Plan for Regulatory Program Review to Support Extended Storage and Transportation of Spent Nuclear Fuel, COMSECY-10-0007, December 6, 2010.

[170] See CRS Report R41908, Energy and Water Development: FY2012 Appropriations, coordinated by Carl E. Behrens.

[171] U.S. NRC, "NRC Appoints Task Force Members and Approves Charter for Review of Agency's Response to Japan Nuclear Event," Press Release No. 11-062, April 1, 2011, http://pbadupws.nrc.gov/docs/ML1109/ML110910479.pdf.

[172] U.S. NRC, Safety Through Defense-in-Depth, July 2011, p. 46. See http://pbadupws.nrc.gov/docs/ML1118/ ML111861807.pdf.

[173] Ibid.

[174] Lawrence Kokajko (U.S. Nuclear Regulatory Commission), Recent NRC Actions Involving Spent Nuclear Fuel Storage, presentation to the Blue Ribbon Commission on America's Nuclear Future, May 13, 2011.

[175] Monitored Retrievable Storage Review Commission, Nuclear Waste: Is There a Need for Federal Interim Storage?, Report of the Monitored Retrievable Storage Review Commission, November 1, 1989.

# U.S. Spent Nuclear Fuel Storage 69

[176] U.S. DOE, Office of Civilian Radioactive Waste Management, Civilian Radioactive Waste Systems: Transportation, Aging and Disposal System Performance Specifications, Rev. 1, DOE/RW-0585, March 2008.

[177] U.S. Nuclear Regulatory Commission, "Consideration of Environmental Impacts of Temporary Storage of Spent Fuel After Cessation of Reactor Operations; Waste Confidence Decision Update; Final Rule," 75 Federal Register 81032, December 23, 2010.

[178] "Dense packing" allows plant operators to store more SNF in a given amount of wet storage pool space by replacing the storage racks. Also, the capacity of an SNF storage pool is generally defined as "full" when there is sufficient remaining capacity to hold one full core load for normal refueling operations and to provide space in case more storage capacity is needed in the event of an accident.

[179] Zhang, Hui (Harvard Belfer Center on Controlling the Atom), Radiological Terrorism: Sabotage of Spent Fuel Pools. INESAP: International Network of Engineers and Scientists Against Proliferation no. 22 (December 2003): 75- 78. Wald, Matthew, New York Times, "Danger of Spent Fuel Outweighs Reactor Threat," March 17, 2011; Frank N. Von Hippel, "It Could Happen Here" (Op-Ed), New York Times, March 23, 2011; and Robert Alvarez, Spent Nuclear Fuel Pools in the U.S.: Reducing the Deadly Risks of Storage, Institute for Policy Studies, May 2011.

[180] Nuclear Regulatory Commission, Spent Fuel Storage in Pools and Dry Casks—Key Points and Questions & Answers, 2011, http://www.nrc.gov/waste/spent-fuel-storage/faqs.html. Another NRC website similarly concludes, "There are two acceptable storage methods for spent fuel after it is removed from the reactor core...." See http://www.nrc.gov/waste/spent-fuel-storage.html.

[181] National Academy of Sciences, National Research Council, Committee on the Safety and Security of Commercial Spent Nuclear Fuel Storage, Safety and Security of Commercial Spent Nuclear fuel Storage: Public Report, 2006, p. 68.

[182] Letter from Nils Diaz, Chairman of U.S. NRC to Senator Pete Domenici; with accompanying report: U.S. Nuclear Regulatory Commission, Report to Congress on the National Academy of Sciences Study on the Safety and Security of Commercial Spent Nuclear Fuel Storage, March 14, 2005.

[183] NRC web page: Spent Fuel Storage in Pools and Dry Casks—Key Points and Questions & Answers, http://www.nrc.gov/waste/spent-fuel-storage/faqs.html.

[184] Wald, Matthew L., "A Safer Nuclear Crypt," New York Times, July 5, 2011.

[185] Jeff Montgomery, "Delaware Energy: PSEG may make small changes for safety," Delaware News Journal, May 18, 2011.

[186] United States Nuclear Waste Technical Review Board, Evaluation of the Technical Basis for Extended Dry Storage and Transportation of Used Nuclear Fuel, December 2010, http://www.nwtrb.gov/reports/eds-final.pdf.

[187] Yoichi Funabashi and Kay Kitazawa, "Fukushima in Review: A Complex Disaster, a Disastrous Response," Bulletin of the Atomic Scientists, March 1, 2012 (this article summarizes a more detailed report published simultaneously in Japanese, by the Rebuild Japan Initiative Foundation); and Institute for Nuclear Power Operations, Special Report on the Nuclear Accident at the Fukushima Daiichi Nuclear Power Station, INPO 11-005, Rev. 0, November 2011.

[188] Robert Alvarez, Spent Nuclear Fuel Pools in the U.S.: Reducing the Deadly Risks of Storage, Institute for Policy Studies, May 2011.

[189] R. W. Borchardt, U.S. NRC, Memorandum to the Commissioners, "Recommended Actions to be Taken Without Delay From the Near-term Task Force Report," SECY-11-124,

September 9, 2011. The staff recommendations were largely accepted by the commission: see Annette L. Vietti-Cook, Memorandum to R. W. Borchardt, "Staff Recommendations— SECY-11-0124, "Recommended Actions to be Taken Without Delay From the Near-term Task Force Report," October 11, 2011.

[190] U.S. NRC, "Issuance of Order to Modify Licenses with Regard to Reliable Spent Fuel Pool Instrumentation," EA-12051, March 12, 2012.

[191] Currently, the inventory of SNF in dry casks is 18,112 MTU. The inventory of SNF, in both wet and dry storage, older than five years is approximately 57,450 MTU (assuming 2,000 MTU fresh SNF generated per year and a total inventory of 67,450 MTU). If any SNF in wet storage currently older than 5 years were transferred to dry storage (approximately 39,338 MTU) were stored in dry casks, it would require a shift of about 80 percent of the current SNF inventory in wet pools. Although these estimates provide a general idea of the quantities involved, these exact amounts would not likely occur because by the time all of the current SNR storage were shifted to dry cask, there would be more aged fuel in storage and more fresh fuel generated.

[192] This estimate of 2,732 new dry casks assumes 50 assemblies per dry cask, based on the 2009 history of 5,567 assemblies loaded into 130 dry casks (43 assemblies per cask), and the 2010 history of 8,606 assemblies transferred to 140 casks (61 assemblies per cask) and 80% of the 170,734 assemblies in wet storage as of the end of 2010. It also assumes, unrealistically, instant transfer, which would actually require several years to accomplish, which could potentially allow for reuse of transport casks. Again, the exact number would depend on the time period over which the shift would occur.

[193] U.S. NRC, "NRC Standard Review Plan for Dry Cask Storage Systems" (NUREG-1536).193 Nuclear Regulatory Commission, "Spent Fuel Storage in Pools and Dry Casks Key Points and Questions & Answers," 2011, http://www.nrc.gov/waste/spent-fuel-storage/faqs.html.

[194] NRC Office of Inspector General, "Audit Report: Audit of NRC's Oversight of Independent Spent Fuel Storage Installations Safety," OIG-11-A-12, May 19, 2011.

[195] U.S. NRC, Memorandum from Stephen D. Dingbaum to R. William Borchardt, Status Of Recommendations: Audit Of NRC's Oversight Of Independent Spent Fuel Storage Installations Safety (OIG-11-A-12), June 30, 2011.

[196] U.S. NRC, Recommendations for Enhancing Reactors Safety in the 21st Century: The Near-Term Task Force Review of Insights from the Fukushima Dai-ichi Accident, July 12, 2011, http://pbadupws.nrc.gov/docs/ML1118/ ML111861807.pdf.

[197] R. W. Borchardt, Executive Director for Operations Policy Issue Memorandum for The Commissioners, "Recommended Actions to be Taken Without Delay from the Near-Term Task Force Report," SECY-11-0124, September 9, 2011; and R. William Borchardt, U.S. NRC, "Briefing on the Japan Near Term Task Force Report— Prioritization of Recommendations," October 11, 2011; and transcript of meeting at http://www. nrc.gov/reading-rm/ doc-collections/commission/tr/2011/20111011.pdf.

[198] See Bunn, Matthew, et al., Interim Storage of Spent Nuclear Fuel: A Safe, Flexible, and Cost-Effective Near Term Approach to Spent Nuclear Fuel Management, Harvard Press, Cambridge, MA, Harvard University Project on Managing the Atom and University of Tokyo Project on Nuclear Energy, 2001, http://belfercenter.ksg.harvard.edu/ files/spentfuel.pdf.

[199] U.S. DOE, "Spent Fuel Storage Requirements 1994-2042," DOE/RW-0431-Rev.1, June 1995.

[200] Letter from Spencer Abraham, Secretary of Energy, to The President, on Site Recommendation (February 14, 2002); U.S. DOE, Recommendation Of The Secretary Of Energy Regarding The Suitability Of The Yucca Mountain Site For A Repository Under

# U.S. Spent Nuclear Fuel Storage 71

The Nuclear Waste Policy Act of 1982 (February 2002), http://www.energy.gov/media/ Secretary_s_Recommendation_Report.pdf.

[201] 42 U.S.C. 10168(d).

[202] Blue Ribbon Commission on America's Nuclear Future, Draft Report to the Secretary of Energy, July 29, 2011, http://brc.gov/sites/default/files/documents /brc_draft_report_ 29jul2011_0.pdf.

[203] Ibid.

[204] Transportation and Storage Subcommittee; DRAFT Report to the Full Commission, Blue Ribbon Commission on America's Nuclear Future, May 31, 2011. See http://brc.gov /sites/ default/files/documents/draft_ts_report_6-1-11.pdf.

[205] Massachusetts Institute of Technology, The Future of the Nuclear Fuel Cycle: An Interdisciplinary MIT Study, MIT, Cambridge, MA, 2011. The MIT study also recommended that "planning for the long term interim storage of spent nuclear fuel—for about a century—should be part of fuel cycle design." Ibid. at page xi.

[206] Bunn, Matthew (Harvard University), testimony before the Blue Ribbon Commission on America's Nuclear Future, May 25, 2010.

[207] This number includes all four DOE sites. The DOE reactors generally shut down in the 1980s, except for a brief restart of the K reactor at the Savannah River Site in 1991.

[208] In addition to these primary agencies responsible for the licensing of the repository, the Department of Transportation is charged with ensuring that waste carriers comply with routing regulations and guidelines, and the Mine Safety and Health Administration of the Department of Labor is responsible for ensuring the health and safety of underground workers at the Yucca Mountain facility.

[209] For more details about the FY2012 budget for DOE, see CRS Report R41908, Energy and Water Development: FY2012 Appropriations, coordinated by Carl E. Behrens.

[210] United States Nuclear Fuel Management Corporation Establishment Act of [2008/2010], S. 3661 in the 110th Congress and S. 3322 in the 111th Congress.

[211] Blue Ribbon Commission on America's Nuclear Future, Draft Report to the Secretary of Energy, January 2012, p. 60.

[212] Massachusetts Institute of Technology, The Future of the Nuclear Fuel Cycle: An Interdisciplinary MIT Study, MIT, Cambridge, MA, 2011.

[213] Blue Ribbon Commission on America's Nuclear Future, Draft Report to the Secretary of Energy, January 2012, p. vii.

[214] Jack Spencer, Heritage Foundation, testimony before the Committee on Science, Space, and Technology's Energy and Environment and Investigations and Oversight Subcommittees, "Review of the Blue Ribbon Commission on America's Nuclear Future Draft Recommendations," October 27, 2011. The comments referred to are essentially identical recommendations of the BRC from its July 2011 draft report.

[215] Douglas Wyatt, Ernie Chaput, and Dean Hoffman, "Siting GNEP at the Savannah River Site: Using Legacy and Infrastructure in a Commercial Energy Park Concept," Waste Management 2008 Conference, 2008, Phoenix, AZ; and Central Savannah River Area Community Team, 2007, Global Nuclear Energy Partnership Siting Study, Final Report, Energy Park on the Savannah River, DE-FG07-06ID14794, April 30, 2007.

[216] See recently William H. Miller, "Used Nuclear Fuel is Good Source of Energy; Reprocessing has Potential for U.S.," Columbia Tribune, December 13, 2011; Clinton Bastin, "We Need to Reprocess Spent Nuclear Fuel, and Can Do it Safely, At Reasonable Cost," 21st Century Science and Technology, Summer 2008; and William Tucker, "There Is No Such Thing as Nuclear Waste," Wall Street Journal, March 13, 2009.

[217] National Academy of Sciences/National Research Council, Board on Radioactive Waste Management, Nuclear Wastes: Technologies for Separations and Transmutation, National Academies Press, Washington, DC, 1996; National Academies/National Research Council, "Committee on Review of DOE's Nuclear Energy Research and Development Program," National Academies Press, Washington, DC, 2008; Frank N. von Hippel, "Nuclear Fuel Recycling: More Trouble Than It's Worth," Scientific American, April 28, 2008; Nuclear Energy Institute, Nuclear Waste Disposal for the Future: The Potential of Reprocessing and Recycling, 2006; Bunn, Matthew, Steve Fetter, John P. Holdren, and Bob van der Zwaan, "The Economics of Reprocessing vs. Direct Disposal of Spent Nuclear Fuel," DOE grant # DEFG26-99FT4028 (Cambridge, MA: Project on Managing the Atom, Harvard University, 2003), December 2003; von Hippel, Frank (Princeton University), "International Impact of U.S. Spent-fuel Policy," prepared statement to the Blue Ribbon Commission on America's Nuclear Future, September 21, 2010, and Nuclear Energy Institute, Policy Brief: Advanced Fuel-Cycle Technologies Hold Promise for Used Fuel Management Program, March 2010, http://www.nei.org/resourcesandstats/documentlibrary/nuclearwastedisposal/policybrief/adv ancedfuelcycle/?page=2; Steve Fetter and Frank N. von Hippel, "Is Reprocessing Worth the Risk?" Arms Control Today, September 2005; Lisbeth Gronlund, David Lochbaum, and Edwin Lyman, Nuclear power in a warming world; Assessing the Risks, Addressing the Challenges, Union of Concerned Scientists, December 2007; Johns Hopkins University, Paul H. Nitze School of Advanced International Studies, Energy, Resources, and Environment Program, Nuclear Fuel Cycle Report, February 2011; and Advanced Nuclear Fuel Cycles—Main Challenges and Strategic Choices, EPRI, Palo Alto, CA: 2010. 1020307.

[218] Massachusetts Institute of Technology, The Future of the Nuclear Fuel Cycle: An Interdisciplinary MIT Study, MIT, Cambridge, MA, 2011, p. xi.

[219] Nuclear Energy Institute, Nuclear Waste Disposal for the Future: The Potential of Reprocessing and Recycling, 2006; and Boston Consulting Group, Economic Assessment Of Used Nuclear Fuel Management In The United States Report (Prepared By The Boston Consulting Group For Areva), July 2006.

[220] Blue Ribbon Commission on America's Nuclear Future, Report to the Secretary of Energy, January 2012, p. 100.

[221] U.S. Congressional Budget Office, Peter R. Orszag, "Costs of Reprocessing Versus Directly Disposing of Spent Nuclear Fuel," Statement before the Senate Committee on Energy and Natural Resources, November 14, 2007, http://www.cbo.gov/sites/default/f iles/cbofiles/ ftpdocs/88xx/doc8808/11-14-nuclearfuel.pdf.

[222] Blue Ribbon Commission on America's Nuclear Future, Report to the Secretary of Energy, January 2012, p. xiv, http://brc.gov/sites/default/files/documents/ brc_finalreport_ jan2012 .pdf.

[223] E-mail from Rodney McCullum, Nuclear Energy Institute, December 11, 2011.

[224] Fettus, Geoff, Tom Cochran, and Christopher Paine, NRDC Comments on Draft Regulatory Basis for Potential Rulemaking on Spent Nuclear Fuel Reprocessing Facilities, Docket ID NRC-2010-0267, July 7, 2011.

[225] See http://www.appropriations.senate.gov/news.cfm?method=news.view&id=eaa626fc-9ba7- 4477-ae48- 25767c9ae814.

[226] Hannah Northey and Nick Juliano, "Bipartisan Senators Begin Legislative Push For Waste Solution," Environment and Energy Daily, April 25, 2012.

In: Spent Nuclear Fuel
Editors: Phillip T. Crawford et al.

ISBN: 978-1-62257-347-9
© 2012 Nova Science Publishers, Inc.

*Chapter 2*

# EXPERIENCE GAINED FROM PROGRAMS TO MANAGE HIGH-LEVEL RADIOACTIVE WASTE AND SPENT NUCLEAR FUEL IN THE UNITED STATES AND OTHER COUNTRIES

## A REPORT TO CONGRESS AND THE SECRETARY OF ENERGY[*]

### *United States Nuclear Waste Technical Review Board*

### ACRONYM LIST

| | |
|---|---|
| AEC | Atomic Energy Commission (United States) |
| AECL | Atomic Energy of Canada Limited |
| ANDRA | National Radioactive Waste Management Agency (France) |
| ASN | Nuclear Safety Authority (France) |
| CEA | Atomic Energy Commission (France) |
| CLI | Local Information Committees (France) |
| CNE | National Review Board (France) |

---

[*] This is an edited, reformatted and augmented version of The United States Nuclear Waste Technical Review Board publication, dated April 2011.

| | |
|---|---|
| CNSC | Canadian Nuclear Safety Commission |
| CoRWM | Committee on Radioactive Waste Management (United Kingdom) |
| DBE | German Service Company for the Construction and Operation of Waste Repositories |
| DOE | Department of Energy (United States) |
| ERDA | Energy Research and Development Authority (United States) |
| HLW | high-level radioactive waste |
| IAEA | International Atomic Energy Agency (United Nations) |
| IRG | Interagency Review Group on Nuclear Waste Management (United States) |
| MUA | multiattribute utility analysis |
| NAGRA | National Cooperative for the Disposal of Radioactive Waste (Switzerland) |
| NAS | National Academy of Sciences (United States) |
| NDA | Nuclear Decommissioning Authority (United Kingdom) |
| NEA | Nuclear Energy Agency (Organization for Economic Cooperation and Development) |
| NRC | Nuclear Regulatory Commission (United States) |
| NWF | Nuclear Waste Fund (United States) |
| NUMO | Nuclear Waste Management Organization (Japan) |
| NWMO | Nuclear Waste Management Organization (Canada) |
| NWPA | Nuclear Waste Policy Act (United States) |
| NWPAA | Nuclear Waste Policy Amendments Act (United States) |
| NWTRB | Nuclear Waste Technical Review Board (United States) |
| OPECST | Parliamentary Office for the Evaluation of Science and Technology Options (France) |
| SKB | Swedish Nuclear Fuel and Waste Management Company |
| SNF | spent nuclear fuel |
| STUK | Radiation and Nuclear Safety Authority (Finland) |
| SSM | Radiation Safety Authority (Sweden) |
| UKAEA | United Kingdom Atomic Energy Authority |
| WIPP | Waste Isolation Pilot Plant (United States) |

UNITED STATES
NUCLEAR WASTE TECHNICAL REVIEW BOARD
2300 Clarendon Boulevard, Suite 1300
Arlington, VA 22201-3367

April 15, 2011

The Honorable John A. Boehner
Speaker of the House
United States House of Representatives
Washington, DC 20515

The Honorable Daniel K. Inouye
President Pro Tempore
United States Senate
Washington, DC 20510

The Honorable Steven Chu
Secretary
U.S. Department of Energy
Washington, DC 20585

Dear Speaker Boehner, Senator Inouye, and Secretary Chu:

The U.S. Nuclear Waste Technical Review Board submits this report, *Experience Gained From Programs to Manage High-Level Radioactive Waste and Spent Nuclear Fuel in the United States and Other Countries*, in accordance with provisions of the 1987 amendments to the Nuclear Waste Policy Act (NWPA), Public Law 100-203, which directs the Board to report its findings and recommendations to Congress and the Secretary of Energy at least two times each year. Congress created the Board to perform ongoing independent evaluation of the technical and scientific validity of activities undertaken by the Secretary of Energy related to implementing the NWPA.

This report explores the efforts of 13 nations to find a permanent solution for isolating high-level radioactive waste (HLW) and spent nuclear fuel (SNF) generated within their borders. It builds on information in the Board's 2009 *Survey of National Programs for Managing High-Level Radioactive Waste and Spent Nuclear Fuel*. Unlike the earlier document, however, this report describes the programs and their histories and discusses inferences that can be drawn from their experiences. We submit the report to provide contextual information for Congress and the Secretary as options are considered for managing HLW and SNF in the United States.

The Board looks forward to continuing to provide useful independent technical and scientific information to Congress and the Secretary that can be used to inform the decision-making process.

Sincerely,

[signed]

B. John Garrick
Chairman

Telephone 703-235-4473   Fax 703-235-4495

## EXECUTIVE SUMMARY

This report explores how 13 nations are carrying out efforts to find a permanent solution for isolating and containing high-level radioactive waste

# 76 United States Nuclear Waste Technical Review Board

(HLW) and spent nuclear fuel (SNF) generated within their borders. Many forces shape how those efforts are designed and implemented. Some of the forces are technical, including choices made about what reactor technology to adopt and about what nuclear fuel cycle to pursue. Others are social and political in nature, including how concerns about intergenerational equity should be addressed and what pace should be followed in implementing a long-term management option. Importantly, the interdependencies, both subtle and overt, between the technical, social, and political forces are inescapable. Because of those interdependencies, what characterizes the national programs most notably is their *variety*.

This report attempts to detail that variety. It builds on the information contained in the U.S. Nuclear Waste Technical Review Board's (NWTRB) *Survey of National Programs for Managing High-Level Radioactive Waste and Spent Nuclear Fuel* (NWTRB 2009). Compared with the earlier document, however, this report is more descriptive and considers the history of national programs.

## Process Considerations

All national programs keenly recognize today that the long-term management of radioactive waste is a complicated *socio-technical* problem, with the social dimension playing an integral role in determining the shape and the ultimate success or failure of a project (see, for example, IAEA 2007). Decisions in all countries about the long-term management of HLW and SNF were initially made by a small group of technical experts, industry representatives, legislative leaders, and government officials. As nuclear power and nuclear waste (and their connection) arose in the mid-1970s as public issues, the need to broaden opportunities for public engagement became clear. Two patterns emerged. First, traditional participatory mechanisms typically were mated with those that were novel and more innovative, such as creating partnerships with potential repository host communities. Second, especially as difficulties arose, many national programs came to recognize the importance of the latter mechanisms and began to rely on them.

Attention has been drawn during the last 20 years to one factor, variously referred to as social or institutional trust that appears to play an important, and perhaps even a decisive, role in determining the effectiveness and perhaps the legitimacy of public-engagement processes (Cvetkovich and Löfstadt 1999; IAEA 2007). Some national programs have come to merit considerable trust and confidence. Other national programs, such as the one in the United States implemented by the U.S. Department of Energy (DOE), either have lost public trust and confidence or seem never to have merited it at all (Carter 1987; Herzik and Mushkatel 1993; SEAB 1993).

In the last decade as well, attention has been drawn to what is presented as a new approach to making choices about the long-term management of HLW and SNF (NEA 2004a; NAS 2003; NWMO 2005). Often referred to as "adaptive management" or "staged decision-making," this approach is actually a refinement of the incremental decision strategy first detailed in the 1950s (Lindblom 1959). At the theoretical level, it is hard to find fault with a decision-making strategy that seems to promise so much. As a more practical matter, however, it is still unclear whether this strategy can be any more successful than earlier efforts in overcoming local and state opposition to specific siting decisions, whether it can be implemented, and even whether it *should* be implemented (Lee 1999).

## DEVELOPMENT, ASSESSMENT, AND ADOPTION OF WASTE-MANAGEMENT OPTIONS

Over the years, national programs have explored a variety of options for the long-term management of HLW and SNF. These options have included the following (IRG 1978):

- Deep-mined geologic disposal
- Burying the waste in deep-sea sediments
- Placing the waste in deep-drilled boreholes
- Partitioning and transmuting the long-lived radioisotopes
- Shooting the waste into space
- Storing the waste indefinitely either above or below ground in a retrievable fashion.

Almost universally, policy-makers have determined that disposal of HLW and SNF in a deep-mined geologic repository is the preferred option for protecting human health and the environment for thousands of years.

In some countries, such as the United States and Germany, that choice was never formally reconsidered. National waste-management programs in other countries, such as France and Canada, also initially chose the geologic disposal option but were compelled by public pressure to evaluate other options explicitly. Although 11 out of the 13 nations considered in this report are officially committed to developing deep-mined geologic repositories as the preferred option for the long-term management of their HLW and SNF, the pace of the development process varies considerably.

## INSTITUTIONAL ARRANGEMENTS FOR EXECUTING WASTE-MANAGEMENT PROGRAMS

Every national waste-management program must address two interrelated questions: Which organizations should be assigned the responsibility for executing which parts of the program, and how will the program be funded? The 13 countries have answered the ques tions in strikingly different ways.

At least two institutions are involved in executing national waste-management programs. The implementer is responsible for developing a safety case, identifying and characterizing candidate sites, and designing, building, and operating the deep-mined geologic repository. The regulator determines whether the approach advanced by the implementer is acceptable. Very early on in some countries, such as the United States, the implementer and the regulator were the same organization. Now there is general agreement that the two institutions should be independent of each other, even if, as in Germany and Japan, they are housed within the same government bureaucracy (see, for example, NEA 2009).

In all 13 national waste-management programs, the regulator is a governmental organization. The current organizational form of the implementer, however, varies considerably across those countries. Some have opted to use a traditional government agency. In other countries, the implementer is a private corporation. The organizational form of the implementer in eight of the 13 the national programs examined in this report has remained the same. For the rest, significant changes often have occurred, typically away from government and towards hybrid or private organizational forms. For many of the significant shifts over time, the root cause appears to be a response to major programmatic challenges.

Both the variety and the evolutions of the implementers' organizational forms seem to demand an inquiry into a question: Which form is best? A simple analysis that tries to associate particular forms with the completion of repository-development milestones produces no clear-cut conclusions: Of the four national programs furthest along, one has been implemented by a government agency (United States), one by a government-owned public service agency (France), and two by private corporations (Finland and Sweden). Advocates of particular forms have supported their choices mostly with impressionistic and unsystematic claims that have not been put to any objective test (Thomas 1993).

Two main approaches toward financing have been adopted by the 13 countries examined in this report. First, special funds have been set up in Canada, Finland, France, Spain, Sweden, Switzerland, and the United States to which waste producers or the consumers of nuclear-generated electricity contribute each year. Second, the expenses incurred by waste-management programs in China, Germany, and the United Kingdom are paid annually out of general government revenues. Although special funds have been established in Belgium, Japan, and the Republic of Korea, the expenses of their current waste-management programs are not covered by the funds but by general governmental revenues.

## TECHNICAL BASIS FOR DEVELOPING DISPOSAL CONCEPTS AND SUPPORTING A SAFETY CASE

In all countries, the implementer has the responsibility for designing a disposal concept that describes a repository system comprising natural and engineered barriers. In most countries, limitations imposed by the geology constrain which disposal concepts can be considered. The implementer typically ends up focusing on one particular geologic formation because of its prevalence or because other formations either are unsuitable technically or cannot be developed because of land-use conflicts. Once a host rock has been chosen, the implementer considers the hydrogeologic environment and determines what, if any, engineered barriers are appropriate as well as how the repository system as a whole will be designed. The implementer is then expected to advance its safety case, a set of arguments and analyses demonstrating why its proposed deep-minded geologic repository will isolate and contain HLW and SNF for as long as society demands. (Various standards and regulatory requirements reflect those demands.) There is broad scientific agreement that deep-mined geologic repositories can be constructed in a wide variety of host-rock formations and hydrogeologic environments, including in salt, crystalline rock such as granite, different clay formations, and unsaturated volcanic tuff.

## SUBSTANCE AND ADOPTION OF HEALTH AND SAFETY STANDARDS AND REGULATIONS

Health and safety standards and regulations serve two purposes. They record society's views about what constitutes acceptable risk, and they establish mechanisms for certifying that an implementer's plan to develop a deep-mined geologic repository can, with a high degree of confidence, satisfy those requirements. All of the 13 national waste-management programs have put in place at least a rudimentary regulatory regime.

Regulators determine what specific standards need to be met. Typically, they must decide on the length of the compliance period, the time over which the repository is expected to satisfy the protective standards. Regulators originally chose compliance periods of several thousand years. The national programs in Belgium, France, Germany, Sweden, Switzerland, and the United States have selected compliance periods of at least 100,000 years and, in most of those cases, as much as 1,000,000 years. In the United Kingdom, the regulator requires the implementer to choose a compliance period and justify its choice. Further, they impose dose constraints or risk limits or some combination of the two. Increasingly national waste-management programs have converged on similar dose constraints and risk limits, at least for the first few thousands of years that a repository is expected to isolate and contain HLW and SNF. Dose constraints vary between 0.1 and 0.3 millisieverts per year. Risk limits vary between a probability of $10^{-5}$ and $10^{-6}$ per year that death or serious health effects will arise over the course of the lifetime of an individual from exposure to radionuclides released from a repository.

Regulators decide how prescriptive their requirements should be. Those choices have produced considerable variation in how much direction the regulators provide. In some national programs, most notably the one in the United States, the rules are quite detailed, laying out specific requirements that the implementer must fulfill in order to get permission to construct or operate a deep-mined geologic repository. Finally, national programs differ in terms of how compliance with the standards is to be demonstrated and what the requirements for demonstration are.

National programs typically have put health and safety regulations into place *after* the implementer has begun to formulate its safety case or to identify candidate sites for a deep-mined geologic repository but *before* specific sites have been chosen. Some interested and affected parties contend that revising regulations *after* a final site already has been selected for a deep-mined geologic repository is inappropriate if the change process is not well explained and supported.

## STRATEGIES FOR IDENTIFYING CANDIDATE SITES FOR A DEEP-MINED GEOLOGIC REPOSITORY

President Jimmy Carter's Interagency Review Group on Nuclear Waste Management (IRG) observed that site-selection strategies for a deep-mined geologic repository necessarily involve passing candidates through what is, in effect, two distinctly different "filters." On the one hand, detailed and quantitative technical requirements have to be met. On the other, sites could be disqualified because of considerations such as the "lack of social acceptance, high population density, or difficulty of access." The two filters could be applied in any order. In the IRG's view, at least, although the suite of sites eventually selected might be different, depending on the order in which the filters were applied, "equally suitable sites should emerge from either approach" (IRG: 1979, 80; 81). Over the years, the United States and other nations have initiated roughly two-dozen efforts to identify or to create processes for identifying potential repository sites. What is noteworthy is how varied those efforts have been.

Part of the variation stems from how the technical filter is constructed. In some cases, efforts to identify candidate sites have focused from the beginning on specific host-rock formations. The choice of those formations has been dictated either by constraints imposed by a country's geology or land-use patterns, by a view that particular host-rock formations possess distinctive advantages in terms of isolating and containing HLW and SNF, or by a combination of these rationales. In other cases, efforts to identify candidate sites cast the net more broadly by enumerating generic qualifying and disqualifying conditions. Qualifying conditions must be satisfied for a candidate site to be considered acceptable; disqualifying conditions eliminate a candidate site from further consideration.

Just as the construction of the technical filter introduces considerable variation in strategies for selecting candidate sites for a deep geologic repository, so does the construction of the nontechnical filter. Arguably this filter's most important property relates to the power that a state or community can exercise. Since the early 1990s, nations outside the United States increasingly have constructed their nontechnical filters in ways that empower local jurisdictions. Many countries begin their site-selection process with a call for communities to volunteer.

The two filters are not independent of each other, except in some theoretical sense. The construction of the nontechnical one may affect the technical one in important ways. To begin with, applying the technical and nontechnical filters is neither purely mechanical nor can it typically be programmed neutrally. Further, implicit in a voluntarist approach is the presumption that a very wide range of geologic features and locations are suitable or can be made suitable. In some cases, this presumption is well-founded. In other cases, even after taking into account fairly general disqualifying conditions, potential disconnects may very well arise, so that applying *both* the technical and nontechnical filters yields a null set of potentially suitable and acceptable sites.

An additional source of variation among national programs can be traced to policies that govern the sequence for accepting or rejecting a candidate site. A country can adopt a "serial" policy whereby sites would be evaluated formally one by one until a suitable site was found. Alternatively, a "parallel" approach can be adopted in which at least two candidate sites would be characterized simultaneously and compared.

## SITE SELECTION FOR A DEEP-MINED GEOLOGIC REPOSITORY

In all national programs, the implementer is responsible for proposing a site to develop as a deep-mined geologic repository. If only one site has been fully characterized at depth, as is the case in the French and American programs, it will be advanced by default if the implementer believes it to be suitable. In most countries, political ratification at the national level of any choice made by the implementer also must take place.

## APPROVAL TO CONSTRUCT A DEEP-MINED GEOLOGIC REPOSITORY

The processes involved in obtaining approval to construct a deep-mined geologic repository are as varied as the processes involved in identifying candidate sites.

In most countries, a representative body, such as the legislature or the Government, makes the final decision. Typically, that body relies on the regulators' advice.

In some countries, however, the regulators make the final determination of whether the proposed repository system complies with established requirements.

## CONCLUSION

In each of the 13 national programs considered in this report, the long-term management of HLW and SNF has proven more complicated and protracted than initially expected. What was formerly viewed as a relatively simple technical task is now recognized as a complex socio-technical problem involving political negotiations and institutional design challenges as well as path-breaking scientific and engineering analyses. Nonetheless, several national programs already have made considerable progress. Sites for a deep-mined geologic repository for HLW and SNF have been selected in four countries—Finland, France, Sweden, and the United States. License applications to construct such a facility have been submitted in two of those nations (the U.S. and Sweden). Applications are likely to be submitted in the other two within the next few years.

The information contained in this report suggests several important conclusions about processes used to develop a deep-mined geologic repository.

- *It is possible to elaborate a disposal concept and to advance a safety case, including quantitative performance assessments, that are widely credible not only to scientific and technical communities but also to broad segments of the general population and political leaders.* It appears as if a deep-mined geologic repository can be developed in a number of different hydrogeologic environments. An open and transparent technical assessment process, including international peer reviews, increases public and political support.

- *It is possible to find communities that are willing to host a deep-mined geologic repository.* From the experience gained in countries where sites have been selected, it appears that some communities do so because of their familiarity with other nuclear activities; others do so because of the economic benefits that might accrue in the future. All of those communities, however, were given a meaningful say in the site-selection process. And all of those communities came to be convinced by the respective implementers that the facility could be constructed and operated safely.

- *Although national programs differ in terms of what is considered an acceptable risk and how to demonstrate whether a deep-mined geologic repository satisfies those standards, international views on these matters are converging.* At least for the first few thousands of years after repository closure, dose constraints across countries are within a factor of three of each other and risk limits are within a factor of ten. Only for compliance periods on the order of 100,000 or 1,000,000 years has no international consensus yet been formed on dose constraints, risk limits, and methodology.

- *Organizational forms differ significantly across countries, but successful ones have several characteristics in common.* Nuclear industry-owned corporations have been successful in Sweden and Finland. A government agency has been successful in France. Rather than organizational form *per se,* what appears to be important are organizational behaviors, such as leadership continuity, funding stability, and the capacity to inspire public trust and confidence over long periods of time.

Today, more than a half-century after electricity was first produced by splitting the atom, the beneficiaries of that energy source have committed themselves to finding ways to manage the radioactive wastes thereby created in a technically defensible and socially acceptable way. That commitment should be a source for optimism, not only for the generation that produced the wastes, but for succeeding generations as well.

# Introduction

The philosopher Hans Jonas has posed what may be the central ethical issue of our age when he tells his readers: "In your present choices, include the future wholeness of Man among the objects of your will" (Jonas 1973, 35-36). This modern-day injunction captures one aspect of the difficult decisions that are required for managing radioactive waste.

Since the early 1950s, more than three-dozen countries have generated electricity by harnessing the energy that holds together the nuclei of heavy elements, such as uranium and plutonium. Although the benefits of the energy produced to societies are substantial, the creation of different radioactive waste streams is an inevitable by-product of these efforts. This report focuses on the experience gained in 13 national programs for managing two of the streams: high-level radioactive waste (HLW) and spent nuclear fuel (SNF).[1]

Initially, the long-term management of HLW and SNF received relatively little attention from national policy-makers (OTA 1985).[2] HLW produced both as a consequence of a country's defense program and as a consequence of reprocessing a country's commercial SNF was stored at first as liquids in tanks. Although it is unclear how the material ultimately would have been managed, this option was considered a satisfactory final solution, especially for waste that had already been produced (AEC 1968 and U.S. Congress 1963).

Perpetual in-tank storage of liquid HLW, however, was soon seen as an unrealistic option. In 1955, the National Academy of Sciences' (NAS) Advisory Committee on Waste Disposal began a study intended to examine other options that might be more effective in isolating and containing HLW over the long-term. Although the writers of the study were careful to note the need for further research, they stated categorically that they "... were convinced that radioactive waste could be disposed of safely in a variety of ways and in a number of sites in the United States." Further, they observed that "... disposal in salt is the most promising method for the near future..." (NAS 1957 3, 6). In retrospect, it appears that the NAS report instilled a sense of complacency in the minds of people dealing with radioactive waste management. The Academy's *imprimatur* left the impression that a "solution" could readily be found once it was needed. And, in fact, nearly a decade would pass before work started to implement the option of disposing of waste in a deep-mined geologic repository. Full development of such a facility to the operational stage has proven to be more problematic than many anticipated.

The difficulties of implementing a plan notwithstanding, the need for satisfactory options for managing HLW and SNF for the long term has not diminished. International and regional organizations, for instance, have highlighted the importance of developing approaches for the long-term management of HLW and SNF. Agreements, such as the *Joint Convention on the Safety of Spent Fuel Management and on the Safety of Radioactive Waste Management,* administered by the International Atomic Energy Agency (IAEA), and legislation, such as the draft directive proposed by the European Commission, call for the near-term adoption of policies and program to address this issue.

In the sections that follow, this report explores how 13 nations are carrying out efforts to find a permanent solution for isolating and containing HLW and SNF generated within their borders.[3] Many forces shape how those efforts are designed and implemented. Some of the forces are technical, including choices made about what reactor technology to adopt and about what nuclear fuel cycle to pursue. Others are social and political in nature, including how concerns about intergenerational equity should be addressed and at what pace the long-term management option should be implemented. Importantly, the interdependencies, both subtle and overt, among the technical, social, and political forces are inescapable. Because of those interactions, what characterizes the national programs most notably is their variety.

This report attempts to describe that variety. It builds on the information contained in the U.S. Nuclear Waste Technical Review Board's (NWTRB) *Survey of National Programs for Managing High-Level Radioactive Waste and Spent Nuclear Fuel* (NWTRB 2009).[4] Unlike the earlier document, this report is more descriptive and considers the history of national programs.[5]

## PROCESS CONSIDERATIONS

All national programs keenly recognize today that the long-term management of radioactive waste is a complicated *socio-technical* problem, with the social dimension playing an integral role in determining the shape and the ultimate success or failure of a project (see, for example, IAEA 2007). National programs increasingly pay attention to issues such as public trust and confidence, transparency, stakeholder engagement, active participation in decision-making processes, and voluntarism.[6] Therefore, it is appropriate that this report begins with a discussion of the experience gained by national programs in devising processes, for it will be those processes that will come into play as countries either make key programmatic decisions in the future or debate whether to reexamine decisions made in the past. In the sections that follow this one, some of these key programmatic decisions are identified and discussed.

## Engagement

Decisions in all countries about the long-term management of HLW and SNF were made initially by a small group of technical experts, industry representatives, legislative leaders, and government officials. When nuclear power and nuclear waste (and their connection) arose in the mid-1970s as issues for the broader public, the need to broaden opportunities for engagement became clear. In response, Belgium, Spain, the United States, Sweden, and Finland were among the first to pass legislation establishing frameworks in which radioactive waste management would be governed. Other countries followed suit. When they started to *implement* the new laws, however, national programs were forced to wrestle with thorny issues (NEA 2004b). Among these were: Should communities, the general public, or representative non-governmental organizations be involved in the decision-making process? And if so, how should they be involved?[7]

Two patterns emerged as these questions were answered. First, traditional participatory mechanisms typically were mated with novel and more innovative ones. Second, especially as difficulties arose, many national programs came to recognize the importance the latter mechanisms and began to rely on them. For example, in the United States, participation frequently takes place under the umbrella of the Administrative Procedures Act and the National Environmental Policy Act. Agencies publish proposals and request public comment on them. Although the agencies are required to explain how they responded to all of the comments, they have broad discretion to accept or reject any of the input received. Yet, as with other nuclear facilities, a highly participatory adjudicatory process has to play out before a deep-mined geologic repository can be licensed. In Canada, the traditional mechanism of holding a public hearing on a proposed option for waste management failed to establish social acceptance for the approach (CEAA 1998). Subsequently, the Nuclear Waste Management Organization (NWMO) intensively and proactively engaged the public to secure acceptance of another waste management option (NWMO 2005).

In Sweden during the 1970s and early 1980s, the Swedish Nuclear Fuel and Waste Management Company (SKB) used an approach to finding a suitable location for a repository that was similar to the one that had been used earlier to identify and propose sites for nuclear power plants. That proved unsuccessful, so starting in the 1990s, SKB encouraged communities to ask tough questions about what was being proposed (Sundqvist 2002). It also established a continuous presence in several communities and strongly supported a so-called "stretching" process in at least one of them.[8] National programs in Belgium and the United Kingdom envision the creation of partnerships between those responsible for implementing national programs for the long-term management of HLW and SNF and the communities that might be willing to host a deep-mined geologic repository (NEA 2010).[9]

It is worthy of note that the various engagement efforts conducted so far have altered the *technical approaches* adopted by national programs in some instances but not in others. For instance, in France, engagement with interested and affected publics highlighted the importance of designing a deep-mined geologic repository so that it would be reversible, thereby triggering a number of technical changes. In contrast, although the Swedish approach to disposing of HLW and SNF, KBS-3, has morphed over time, none of the technical changes were the result of input from communities that might serve as hosts for a deep-mined geologic repository. Instead, serious engagement seems to have served other functions, one of which is to increase public trust and confidence both in the local communities and at the national level.

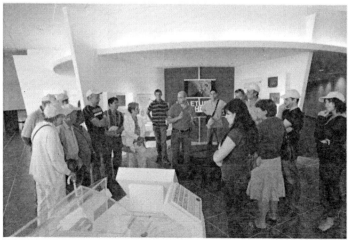

Source: ANDRA.
ANDRA personnel explain the French repository design to members of the public.

## Public Trust and Confidence

Attention has been drawn during the last twenty years to one factor, variously referred to as social or institutional trust, which appears to play an important, and perhaps even a decisive, role in determining the effectiveness, and perhaps the legitimacy, of public-engagement processes (Cvetkovich and Löfstadt 1999; IAEA 2007). Some national programs have come to merit considerable trust and confidence. In Sweden, for instance, the nuclear regulators enjoy broadly based social trust. Communities considering whether to host a deep-mined geologic repository express confidence in the regulators' judgments and are willing to base their decisions on those judgments (Åhagen et al. 2006).

The national program in Finland has not had to confront a deficit of social trust. Other national programs, however, such as the one in the United States implemented by DOE, either have lost public trust and confidence or seem never to have merited it at all (Carter 1987; Herzlik and Mushkatel 1993). In France, a lack of social trust led to a moratorium on site-selection in 1990 (Barth 2009).

So critical is the development of social trust that several national programs have explored the issue specifically and, as indicated by reference to published reports below, have concluded that, absent widespread social trust, efforts to implement a long-term radioactive waste management approach are likely to confront significant, if not overwhelming, obstacles. In the United States, a Task Force advising the Secretary of Energy observed (SEAB 1993, 39):

> The legacy of distrust created by the Department's history and culture will continue for a long time to color public reaction to its radioactive waste management efforts. Only a sustained commitment by successive Secretaries of Energy can overcome it.

In Canada, a report by NWMO reached essentially the same conclusion (NWMO 2005, 75):

> Many examples have been brought forward of incidents in which the industry and/or government have acted in what is perceived to be a self-interested and secretive manner. For these participants, this is a key area in which trust must be built before proceeding with any approach for the long term management of used nuclear fuel.

Similarly in the United Kingdom, an advisory group appointed by Government, the Committee on Radioactive Waste Management (CoRWM), noted (CoRWM 2006, 14):

> There is also a high degree of historical distrust of the nuclear industry and those charged with developing waste management facilities, which has led to a breakdown of previous attempts to implement a policy.

In Canada and the United Kingdom, where waste-management programs are attempting to rebuild social trust, the initial signs seem positive. Sustaining trust appears to be a key consideration in decisions now being made. Research has shown, however, that recovering trust once it has been lost can be challenging (see, for example, Kasperson et al. 1992). Organizations that are not trusted have little "slack." What might have been forgiven were the agency considered trustworthy is often viewed as compelling evidence that the organization has not changed its ways fundamentally. Moreover, the context in which choices are made can make a difference. As a study by the IAEA recognizes, the situation changes once national programs go beyond their generic stages and confront the question of selecting candidate sites for a deep-mined geologic repository. Social trust, which might be established early on, may be harder to maintain as the programs mature or site-specific proposals are made (IAEA 2007).[10]

## Decision-Making Strategies

In the last decade, attention has been drawn to what is presented as a new approach to making choices about the long-term management of HLW and SNF (NEA 2004a; NAS 2003; NWMO 2005). Often referred to as "adaptive management" or "staged decision-making," this approach is actually a refinement of the incremental decision strategy first detailed in the 1950s (Lindblom: 1959). Rather than decisive choices being made at the front end of the process, decisions are made in a stepwise fashion. At predetermined decision points, the work of national waste-management programs is reviewed and evaluated systematically by all interested and affected parties, and an explicit choice is made about whether to proceed along the program's proposed path or to reconsider what is to be done. A premium is placed on "organizational learning, flexibility, reversibility, auditability, transparency, integrity, and responsiveness" (NAS 2003, 124). National programs in

Canada, France, Japan, Sweden, and Switzerland explicitly make this approach a centerpiece of their strategy for the long-term management of HLW and SNF. Other programs do not reject this approach but view it as nothing new, maintaining that they already have incorporated the strategy at least implicitly. [11] At the theoretical level, it is hard to find fault with a decision-making strategy that seems to promise so much. As a more practical matter, however, it is still unclear whether it can be any more successful than earlier efforts in overcoming local and state opposition to specific siting decisions, whether it can be implemented, and even whether it *should* be implemented (Lee 1999). The many tensions between theory and practice are well understood and often are acknowledged by the strategy's advocates, but two of them bear repeating here.

- In theory, supporters of a staged strategy contend that it fosters public acceptance by making national programs more responsive, even to the point of abandoning a project after substantial investments in it have been made. In practice, however, the approach detailed by Lindblom works best when there is *fundamental agreement* on outcomes at the start (Thompson and Tuden 1959).
- In theory, under a staged approach, those responsible for carrying out national waste- management programs openly acknowledge errors and uncertainties and make adjustments to correct mistakes. In practice, it is not unheard of that organizations cover up errors rather than acknowledge them. But even assuming a completely open-minded bureaucracy, discovering and rectifying mistakes is especially difficult when it comes to evaluating the performance of complex systems such as a deep-mined geologic repository (Vaughn 1996).

Perhaps the most important contribution that the proponents of such a structured and staged strategy have made is to identify a set of aspirations and objectives that national programs should strive to meet. It is by no means clear that a staged decision-making strategy is the only, or even the best, way to do so.

## Remaining Questions

Although there seems to be a general consensus about the importance of process considerations in the long-term management of HLW and SNF, questions remain about how to design processes that are effective given each nation's political culture. For example,

- Why will some communities volunteer to host a deep-mined geologic repository while others in the same country will not?
- Although adaptive staged management may be the most promising decision-making strategies for the long-term management of HLW and SNF, it is unclear how to design an approach that that can address issues such as possible disconnects between the need for mid-course corrections when evidence suggests that they may be required and the capability of institutions to make those changes.

There is no one recipe that can be chosen to ensure a successful process. Better understanding of processes, however, may lead to improved approaches (e.g., Chilvers and Burgess 2008; Cotton 2009) and adjustments also may be needed through time (e.g., Krutlia et al. 2010). Research on such issues has not been extensive, and much more knowledge is needed. The words of Jacob (1990, 164) remain true today: "While vast resources have been expended on developing complex and sophisticated technologies, the equally sophisticated political processes and institutions required to develop a credible and legitimate strategy for nuclear waste management have not yet been developed." More research also is needed on beliefs and perceptions and their linkage to behavior. For example, open questions remain about precautionary attitudes (e.g., Sjoberg 2009) and the connections among such factors as economic benefits, risk perception, and trust (e.g., Chung and Kim 2009).

## Major Summary Points

- Two clear patterns describe how the participatory process has evolved in many nations. First, traditional mechanisms for public participation have been supplemented by novel and more innovative ones. Second, the more difficulty a program has encountered in the past, the more that national programs have come to recognize the importance of and to rely on active mechanisms for public engagement.
- Most existing siting processes now being implemented or contemplated by national waste-management programs involve some delegation of decision-making power to interested and affected outside parties.

- Although increased public involvement has affected key technical choices made by some national waste-management programs, it has had little effect on others.
- Some programs have undertaken significant actions to rebuild public trust and confidence, but, in many of those cases, the programs have not yet been put to the test; it remains to be seen how successful national programs will be in recovering lost trust.
- A staged decision-making strategy may offer the most promise for developing a deep-mined geologic repository, but it is unclear how well it can be implemented.

## DEVELOPMENT, ASSESSMENT, AND ADOPTION OF WASTE-MANAGEMENT OPTIONS

Over the years, national programs have explored a variety of options for the longterm management of HLW and SNF. These options have included the following (IRG 1978):

- Deep-mined geologic disposal
- Burying the waste in deep-sea sediments
- Placing the waste in deep-drilled boreholes
- Partitioning and transmuting the long-lived radioisotopes
- Shooting the waste into space
- Storing the waste indefinitely either above or below ground in a retrievable fashion.

Almost universally, policy-makers have determined that disposal of HLW and SNF in a deep-mined geologic repository is the preferred option for protecting human health and the environment for thousands of years.[12] As one international group put it, the option "... provides a unique level and duration of protection... It takes advantage of the capabilities of both the local geology and the engineered materials to fulfill specific safety functions in a complementary fashion, providing multiple and diverse barriers..." (NEA 2008b 7, 14).

In virtually all countries, the choice of disposal in a deep-mined geologic repository originally was reached implicitly by accepting a technical consensus that began to form more than 50 years ago with the publication of the NAS report.

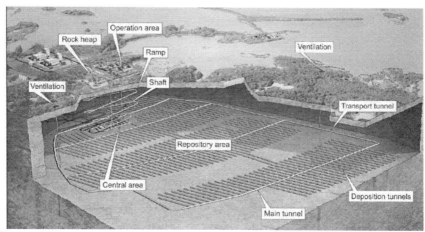

Source: SKB.
Conceptual view of the Swedish deep-mined geologic repository.

In some countries, that choice was never formally reconsidered.[13] For example, by 1977, the first so-called KBS report suggested that vitrified HLW should be placed in a deep-mined geologic repository. That report, however, also outlined the possibility of depositing SNF in such a facility. As Sweden gradually shifted to a strategy of direct disposal of SNF, SKB developed two conceptual plans culminating in the adoption of KBS-3 (SKB 1978a, 1978b, 1983). During that period, some details were finalized: The copper option was chosen for the waste package instead of ceramic; vertical emplacement of the waste packages was picked over horizontal emplacement; and bentonite rather than a mixture of bentonite and quartz sand was selected as the buffer. But one element of the conceptual plans has remained constant: A deep-mined geologic repository would be developed in the crystalline rock that is so prevalent in Sweden.

The Germans began in 1965 to construct an underground laboratory at Asse in Lower Saxony to determine whether salt might be a suitable host formation for disposing of HLW. By the early-1970s, the federal government selected that rock type.

In 1973, a siting process was launched for the Integrated Waste Management Center, where a repository for HLW and long-lived intermediate-level waste would be co-located with a reprocessing facility. A site close to the community of Gorleben, also in Lower Saxony, was chosen four years later (Hocke and Renn 2010).[14]

In the United States between 1965 and 1967, researchers from Oak Ridge National Laboratory carried out studies at the inactive Carey Salt Mine outside of Lyons, Kansas. Their work helped confirm the optimism expressed in the NAS report about the suitability of salt as a host rock. The Atomic Energy Commission (AEC) in 1970 officially selected disposal in a deep-mined geologic repository as the sole option for the long-term management of HLW (AEC 1970). Disposal in a deep-mined geologic repository was more firmly established in law with the passage of the Nuclear Waste Policy Act (NWPA) in 1982.

National waste-management programs in Canada and France also initially chose the geologic disposal option but were compelled by public pressure to reconsider other options explicitly. In 1978, the Governments of Canada and Ontario announced the creation of a program to dispose of radioactive wastes from nuclear power reactors in a deep-mined geologic repository developed in intrusive igneous rock.

To establish the technical basis for this program, the Whiteshell underground laboratory was built near the town of Pinawa in Manitoba, and experiments were conducted there for more than a decade.

In 1994, Atomic Energy of Canada Limited (AECL) submitted a comprehensive Environmental Impact Statement based on the concept of placing SNF in corrosion-resistant copper containers at a depth of between 500 and 1,000 meters in plutonic rock located in the Canadian Shield (AECL 1994). After extensive public hearings, the Government-appointed Seaborn Panel issued its findings evaluating the concept (CEAA 1998).

- Broad public support is necessary in Canada to ensure the acceptability of a concept for managing nuclear fuel wastes.
- Safety is a key part, but only one part, of acceptability. Safety must be viewed from two complementary perspectives: technical and social.
- From a technical perspective, the safety of the AECL concept has been on balance adequately demonstrated for a conceptual stage of development, but from a social perspective, it has not.
- As it stands, the AECL concept for deep geological disposal has not been demonstrated to have broad public support. The concept in its current form does not have the required level of acceptability to be adopted as Canada's approach for managing nuclear fuel wastes.

When the Canadian Government subsequently accepted the findings and recommendations of the Seaborn Panel, in effect it threw out the waste-management option that had been the basis of Canada's waste-management program for 20 years.

Starting from scratch NWMO reviewed more than a dozen waste-management options and decided to analyze three of them in-depth:

- Deep geologic disposal in the Canadian Shield
- Storage at nuclear reactor sites
- Centralized storage above or below ground.

In the end, NWMO recommended to Government a fourth option, what it termed "a technical method and a management system" (NWMO 2005). This option had the following features:

- Ultimate centralized containment and isolation of used nuclear fuel in an appropriate geologic formation
- Phased and adaptive decision-making
- Optional shallow storage at the central storage site prior to placement in the repository
- Continuous monitoring
- Provision for retrievability
- Citizen engagement.

In 2007, Government adopted NWMO's recommendations and accepted the Adaptive Phased Management approach as the best option for the long-term management of nuclear-fuel waste.

The French waste-management program underwent a similar crisis in public confidence and also was forced to re-examine its choice of waste-management options. By the mid-1980s, the Government had identified four specific sites—in clay, crystalline rock, schist, and salt—where a repository might be developed. Some test boreholes were drilled, and analyses were undertaken to determine how to optimize the waste isolation system, "... taking into account the overall characteristics of the particular host rock" (Carter 1987, 323). The four sites were selected based on technical judgments on their merit, and their selection did not involve much participation either by the public at large or by the local authorities. Subsequently, intense local opposition emerged at all four sites, and the Prime Minister called a halt to the site-selection process in February 1990.

Parliament passed new legislation governing radioactive waste in December 1991. The Research in Radioactive Waste Management Act reopened the question of what waste-management option should be adopted. It laid out in clear terms major areas of research, which were to be carried out by the National Radioactive Waste Management Agency (ANDRA). They were:[15]

- Partition and transmutation
- Waste packaging and effects of long-term surface storage

The feasibility of reversible and non-reversible deep-mined geologic disposal through studies conducted in underground research laboratories.

The Parliamentary Office for the Evaluation of Science and Technology Options (OPECST) held three days of public hearings on radioactive waste management in early 2005. The hearings covered partitioning and transmutation, deep-mined geologic disposal, and longterm interim storage. The hearings were conducted to inform a report that was prepared by OPECST and released later that year (OPECST 2005).

Following the national public debate, and taking it into account, the Sustainable Management of Radioactive Materials and Waste Act was submitted to and passed by Parliament in 2006. The Act selected phased deep geologic disposal as the preferred option for managing HLW and dictated that a deep-mined geologic repository be developed at a site to be chosen by 2015. However, in anticipation of the deployment of fast reactors, the research then under way into partitioning and transmutation of long-lived radioactive elements would be continued.

The situation in the United Kingdom is a bit different and more complicated. That country envisioned reprocessing all of its SNF. Just as in the United States, some thought was given to storing the liquid HLW in tanks for an indefinite period. That waste-management option, however, was abandoned in 1976 when a Royal Commission on Environmental Pollution severely criticized it. By then, Government concluded that developing a repository for vitrified HLW was too onerous and decided to concentrate on managing the long-lived intermediate-level waste from reprocessing. Over the next few years, attempts were made to obtain from local communities "planning permission" to drill surface boreholes to determine whether sites selected by the Institute of Geological Sciences might be suitable for developing a deep-mined geologic repository. Planning permission was granted at one of the sites, and a Public Inquiry was launched at another one. By 1981, Government

98      United States Nuclear Waste Technical Review Board

had come to view developing a deep-mined geologic repository as politically infeasible and had concluded that long-term surface storage of solidified HLW, not disposal, should become the nation's preferred waste-management option.

Not until the late 1980s did the United Kingdom venture back into the realm of geologic disposal of HLW and SNF. Industry-owned UK Nirex Limited applied for planning permission to construct a Rock Characterization Facility near the reprocessing site at Sellafield in West Cumbria. The facility would have permitted characterization of the rocks to determine whether they are suitable for a deep-mined geologic repository for long-lived intermediate-level waste. After a Public Inquiry at which strong opposition to the facility was voiced, the local government refused planning permission. This decision was confirmed at the national level in 1997.

Unlike the similar situation two decades previously, Government realized that it could not avoid addressing the question of long-term management of HLW and SNF. To do so, it launched the Managing Radioactive Waste Safely consultation program in 2001. As part of the program, CoRWM was asked in 2003 to make recommendations on what options the country should adopt.

CoRWM began its evaluation by identifying waste management options that had been suggested in the past. The list included the following:

- Interim or indefinite storage on or below the surface
- Near-surface disposal, a few meters or tens of meters below ground
- Deep disposal, with the surrounding geology providing a further barrier
- Phased deep disposal, with storage and monitoring for a period
- Direct injection of liquid wastes into rock strata
- Disposal at sea
- Subseabed disposal
- Disposal in ice sheets
- Disposal in subduction zones
- Disposal in space, into high orbit, or propelled into the Sun
- Dilution and dispersal of radioactivity in the environment
- Partitioning of wastes and transmutation of radionuclides
- Burning of plutonium and uranium in reactors

CoRWM then developed a set of screening criteria to narrow the list. It used two rounds of public-engagement meetings to inform its assessment of the options. The shortlist for detailed evaluation contained only three options, which, in CoRWM's view, could be implemented in principle to manage the country's entire inventory of HLW and SNF: interim storage, geological disposal, and phased geological disposal. In 2006, CoRWM made its recommendations (CoRWM 2006). Several months later, Government responded by accepting geological disposal coupled with safe and secure interim storage as the way forward for the long-term management of the United Kingdom's higher activity wastes and designated the Nuclear Decommissioning Authority (NDA) as the responsible organization (UKG 2006).

Although 11 out of the 13 nations considered in this report are officially committed to developing deep-mined geologic repositories as the preferred option for the long-term management of their HLW and SNF, the pace of the development process varies considerably. For some national waste-management programs, moving forward expeditiously is the best way to address concerns about intergenerational equity. Those countries typically have made firm decisions about whether to reprocess their SNF. Nations projecting a longer time horizon believe that many of their citizens still have not accepted that deep-mined geologic repositories are an appropriate strategy for permanent disposal of HLW and SNF and that they therefore must plan for an extended period prior to development of a repository during which technical work and public engagement will continue (Table 1).

## Major Summary Points

- A broad consensus has emerged over the last 50 years that properly sited and designed deep-mined geologic repositories can isolate and contain HLW and SNF for many thousands of years, thereby adequately protecting human health and the environment.
- Countries whose waste-management programs have been challenged and disrupted have subsequently conducted an explicit and transparent evaluation of options to inform a national decision on whether to adopt disposal in deep-mined geologic repositories as the preferred approach. In all cases, that option was selected.
- The pace at which national repositories are being developed varies considerably. Factors affecting the development schedule include public and political acceptance, attitudes toward nuclear power, views on intergenerational equity, and a desire not to foreclose the possibility of extracting additional energy from SNF.

# Table 1. Status of the repository-development programs for HLW and SNF

| COUNTRY | STATUS |
| --- | --- |
| United States | License application for a repository at Yucca Mountain in Nevada was submitted to the Nuclear Regulatory Commission (NRC) in June, 2008 Subsequently, the Administration sought to withdraw the application and to defund the project No final decisions have been made |
| Belgium | Geologic disposal has not been officially adopted as the country's preferred long-term waste management option, although investigations have been conducted in clay Start of repository operations is anticipated in the 2040 time frame |
| Canada | Engagement with the public determined the criteria that will be used to select candidate sites for a repository Eight communities in two Provinces have expressed an interest in learning more about the possibility for hosting a repository No date has been for the start of repository operations |
| China | Preliminary site investigations are under way at Beishan in the Gobi Desert Start of repository operations is anticipated in the 2050 time frame |
| Finland | A site at Olkiluoto near the community of Eurajoki has been selected for a repository Site investigations and construction are under way Start of repository operations is anticipated in the 2020 time frame |
| France | A site in the Meuse/Haute-Marne region has been selected for a repository Site investigations are underway Authorization to construct the repository is anticipated in 2017 Start of repository operations is anticipated in the 2025 time frame |
| Germany | After a hiatus of 10 years, Government has announced that site investigations at Gorleben would resume No date for the start of repository operations has been set |
| Japan | Generic investigations have been conducted in crystalline rock In 2002, the Government sought volunteer communities to explore the feasibility of constructing a final repository for HLW To date, no community has agreed to volunteer No date has been set for the start of repository operations |

| COUNTRY | STATUS |
|---|---|
| Republic of Korea | The repository development process for HLW and SNF has not begun |
| Spain | Geologic disposal has not been officially adopted as the country's preferred option for long-term waste-management |
| Sweden | A site in the municipality of Östhammar has been selected based on site investigations A license application to construct a repository was submitted in March 2011 Start of repository operations is anticipated in the 2023 time frame |
| Switzerland | The first of three phases of the site-selection process has begun Start of repository operations is anticipated no sooner than 2040 |
| United Kingdom | Two communities in West Cumbria are discussing with Government the possibility of participating formally in the site-selection process for a repository Start of repository operations is anticipated in the 2040 time frame |

## INSTITUTIONAL ARRANGEMENTS FOR EXECUTING WASTE-MANAGEMENT PROGRAMS

Every national waste-management program must address two interrelated questions: Which organizations should be assigned the responsibility for executing which parts of the program, and how will the program be funded? The 13 countries have answered those questions in strikingly different ways.[16]

### Organizations

At least two institutions are involved in executing national waste-management programs: the implementer and the regulator. The implementer is responsible for developing a safety case, identifying and characterizing candidate sites, and designing, building, and operating the deep-mined geologic repository. The current organizational form of the implementer varies considerably across the countries.

Some nations have opted to use a traditional government agency. In the United States, at the same time that geologic disposal was selected as the preferred waste-management option, the AEC also deliberated over what type of organization should develop and operate a deep-mined geologic repository.

102     United States Nuclear Waste Technical Review Board

Only the tiny Division of Industrial Participation voiced any objections to the decision that the responsibilities should belong to the federal government.[17] In Germany and the Republic of Korea as well, the implementer of the waste-management program is a government agency.[18] France considers its implementer to be a government-owned public service agency. In the United Kingdom, the NDA, a non-departmental public body under the purview of the Department of Energy and Climate Change, is the implementer.

In other countries, the implementer is a private corporation. Beginning in 1973, SKB, jointly owned by the nuclear power producers, took charge of efforts to develop an approach for disposing of, at first, HLW, and now, SNF.[19] In Canada and in Finland, the implementers, NWMO and Posiva Oy respectively, are organized and tasked in much the same way.[20] In still other countries, the implementers are hybrid organizations. Government-owned corporations are the implementers in Belgium, China, and Spain. In two nations, the implementing organization takes on idiosyncratic forms, combining in varying degrees public and private characteristics. In Japan, a private non-profit corporation, the Nuclear Waste Management Organization (NUMO), was established by the owners of nuclear power plants and is supervised by the Ministry of Economy, Trade, and Industry. In Switzerland, a private/public consortium of radioactive waste producers, the National Cooperative for the Disposal of Radioactive Waste (NAGRA), includes the owners of nuclear power plants and the Federal State.

The organizational form of the implementer in seven of the 13 the national programs examined in this report has remained the same (Belgium, Finland, France, Germany, Sweden, Switzerland, and the United States). In France, for example, the Atomic Energy Commission (CEA) was responsible for the first research undertaken to develop a deep-mined geologic repository. Because of concerns about waste management efforts being overwhelmed within the large CEA, ANDRA was established in 1979 as a separate unit within the CEA. As will be discussed below, ANDRA encountered difficulties when it tried to initiate site investigations. That experience led in 1991 to Parliament's passage of the Research in Radioactive Waste Management Act, one provision of which called for ANDRA to be removed from the CEA and made into a government-owned public service agency.

In Canada, Japan, and the United Kingdom, government agencies were the initial implementers, perhaps the result of a legacy dating from the time when nuclear energy policy was deemed so exceptional that it had to remain under tight government control. Significant shifts over time, however, have taken place, typically away from government and towards hybrid or private organizational forms. These changes appear to have a common root cause, namely, a response to major programmatic challenges.

AECL was established in 1950 as a federal Crown corporation with the responsibility for managing Canada's nuclear energy program (including radioactive waste management), conducting research and development, and carrying out a number of commercial operations, such as the promotion of the CANDU reactor. The rejection of AECL's conceptual disposal plan in 1998 led Government to revise the structure of its waste-management program. The subsequent passage of the Nuclear Fuel Waste Act in 2002 transferred responsibility to NWMO, which is jointly owned by the three nuclear utilities that have generated SNF.

The Japanese Government first placed efforts to develop a deep-mined geologic repository in the hands of the country's Atomic Energy Commission. But the waste-management program developed slowly. Concerned that the absence of a repository could impede the construction of additional nuclear power plants, Parliament passed the Final Disposal of Specific Radioactive Wastes Act in 2000, which established NUMO.

In the United Kingdom, responsibility for developing a deep-mined geologic repository was initially given to the Atomic Energy Authority (UKAEA). Its inability to get planning permission from a number of communities to characterize sites in the mid-to-late 1970s led to the creation of the Nuclear Industry Radioactive Waste Executive and then to UK Nirex. As noted previously, the failed attempt in the late 1990s to obtain planning permission for the Sellafield Rock Characterization Facility forced Government to reconsider its waste-management efforts, culminating in the adoption of the Managing Radioactive Waste Safely program. In 2005, ownership of UK Nirex was transferred from the nuclear industry to the government of the United Kingdom and, in 2007, UK Nirex was disbanded and its staff transferred to the NDA. Within the NDA, the Radioactive Waste Management Directorate has primary responsibility for developing a deep-mined geologic repository. Its conversion into a "site licensing company" is under way.

Both the variety and the evolutions of the implementers' organizational form seem to demand an inquiry into the question: Which form is best? A simple analysis that tries to associate particular forms with the completion of repository-development milestones produces no clear-cut conclusions: Of the four national programs furthest along, one has been implemented by a government agency (United States), one by a government-owned public service agency (France), and two by private corporations (Finland and Sweden).

104 United States Nuclear Waste Technical Review Board

In 1982, when Congress passed the NWPA, it also commissioned a study to identify an appropriate organizational form for the American implementer. The Advisory Panel on Alternative Means for Financing and Managing Radioactive Waste Facilities, among other things, evaluated four arrangements (AM FM: 1984). It ended up endorsing a FEDCORP, that is, a government-owned corporation structured to operate more like a private enterprise than a government agency. The panelists reached that conclusion based on the weights they gave to the lengthy lists of pros and cons associated with each arrangement. Both DOE and the U.S. nuclear industry, using their own weights, held that the *status quo* should be maintained (see, for example, Harrington 1985).[21] In 2001, Congress asked the DOE to follow up on the earlier evaluation. DOE evaluated three arrangements and developed a lengthy list of pros and cons associated with each arrangement. Ultimately, DOE was not prepared to choose among the alternatives and concluded that they should be studied further once a decision on the Yucca Mountain site had been made (DOE 2001).

Combining the experience gained by 13 national programs with the more detailed analyses undertaken in connection with the U.S. program suggests that the answer to the question posed above is that it depends. Some interested and affected parties will care strongly that the implementer is responsive to their views; others will want the implementer to focus on meeting schedules or minimizing expenditures. Each country seems to find its own particular way to resolve these value-based conflicts. In the final analysis, what may be most critical is not organizational form *per se* but organizational behaviors.

In all 13 national waste-management programs, the regulator is a government organization. The regulator determines whether the approach advanced by the implementer is acceptable. Very early on in some countries such as the United States, the implementer and the regulator were the same organization. Now there is general agreement that the two institutions should be independent of each other, even if, as in Germany and Japan, they are housed within the same government bureaucracy (see, for example, NEA 2009). In some countries, such as the United Kingdom and the United States, multiple regulators have authority over the development of deep-mined geologic repositories.

In six of the 13 nations, independent oversight organizations have been created. Some of them, including the NWTRB, the National Review Board in France (CNE), the Nuclear Waste Management Commission in Germany, and the Nuclear Safety Commission in Switzerland, limit their oversight to technical matters. Others, including the National Council for Nuclear Waste in

Sweden and the reconstituted CoRWM in the United Kingdom, have a broader mandate and provide oversight on non-technical matters as well as technical ones. All of these oversight organizations reach conclusions about how well the implementer is carrying out its responsibilities and make recommendations for improvement to some government body or bodies. Typically, they lack the power to enforce their recommendations, although in some countries the presumption is that their advice will be accepted (see, for example, U.S. Congress 1987).

## Finances

Developing and operating a deep-mined geologic repository is a decades-long undertaking. That period can stretch out even further if a nation decides to close the facility only after an extended monitoring program. Ensuring adequate funding for such a lengthy program can present substantial challenges. Credible cost estimates have to be calculated and then periodically updated as new information emerges. Mechanisms have to be put into place to collect the needed revenue, to make certain that it is spent only for its intended purpose, and to undertake long-term planning to develop a deep-mined geologic repository.[22]

Two main approaches toward financing have been adopted by the 13 countries examined in this report. First, special funds have been set up in Canada, Finland, France, Spain, Sweden, Switzerland, and the United States to which waste producers or the consumers of nuclear-generated electricity contribute each year. Second, the expenses incurred by waste-management programs in China, Germany, and the United Kingdom are paid annually out of general government revenues. Although special funds have been established in Belgium, Japan, and the Republic of Korea, the expenses of their current waste-management programs are not covered by the funds but by general government revenues.[23]

Although the total system life-cycle costs of developing, operating, and decommissioning a deep-mined geologic repository are estimated by a government agency in all countries but Finland and France, national waste-management programs that have established special funds have done so in different ways. In Canada, NWMO proposed a rather complex formula for determining the annual contributions from the three nuclear utilities and AECL (NWMO 2007). Government approved the formula in 2009. Each contributor deposits its payments in separate, segregated trust funds. Approximately $60

106     United States Nuclear Waste Technical Review Board

million (Canadian) is collected annually. Each year in Finland, generators pay the difference between the estimated fund target liability incurred to date and the amount that they had previously paid (Finnish Energy Industries 2007). Payments can be in securities. Excess payments can be recovered. The fund is controlled and administered by the Ministry of Trade and Industry.

Nuclear waste generators in France establish reserves to cover the full costs of long-term management of their wastes (ANDRA 2009). These reserves are held separately within each company. Every three years, each generator transmits to an independent government commission a report describing how the costs were estimated and the choices adopted for the composition and management of the reserves. The commission, composed of parliamentarians and individuals appointed by Parliament and the Government, has the authority to require additional contributions if it concludes that a generator's payments are insufficient. In Sweden, the holders of a license to operate a facility that generates radioactive waste must pay a fee to the State.

The fee, which is now set by Government every third year, is calculated on a per-kilowatt-hour basis. Owners of facilities that no longer operate must pay special fees for the uncovered costs of managing their SNF. If the fund is insufficient, the generators must provide a guarantee to make up any deficit (Kärnavfallsfonden 2010). The accumulated funds are managed by a government authority.

In Switzerland, approximately 35 percent of the anticipated total system life-cycle costs already has been paid for and spent. Another 17 percent is the costs to be incurred by the generators between 2008 and the time that their facilities are decommissioned. The remaining 48 percent will be covered by contributions to the fund (SFOE 2009).

In the United States, DOE is required to prepare an annual estimate of the total system life-cycle costs of developing, operating, and closing a deep-mined geologic repository for SNF and HLW. Based on that estimate, it has to recommend to Congress whether the legislatively dictated fee of 1 mil/kilowatt-hour[24] of nuclear-generated electricity that utilities contribute to the Nuclear Waste Fund (NWF) should be revised.

Although called a trust fund, the monies held in the NWF as a practical matter are not segregated. Thus they effectively pay for other government programs. Moreover, because Congress appropriates money to develop a repository strictly on an annual basis, DOE has to cope with widely varying budgets and often does not have direct access to the funds that have accumulated.

## Major Summary Points

- Implementers of waste-management programs take on a variety of institutional forms. In general, there does not seem to be any connection between institutional form and "progress" toward constructing and operating a deep-mined geologic repository.
- The choice of organizational form for the implementer depends in each country on how value-based conflicts are resolved. There does not seem to be "one best way" that can be universally applied.
- The costs of some national programs are paid from general government revenues. Other national programs have devised formulas and procedures for determining how much money should be collected from waste generators based on estimates of the total system life-cycle cost. In most countries, those funds are deposited into segregated accounts.

## TECHNICAL BASIS FOR DEVELOPING DISPOSAL CONCEPTS AND SUPPORTING A SAFETY CASE

In all countries, the implementer has the responsibility for developing a disposal concept that describes a repository system comprising natural and engineered barriers. In most countries, limitations imposed by the geology constrain which disposal concepts can be considered. The implementer typically ends up focusing on one particular geologic formation because of its prevalence or because other formations either are unsuitable technically or cannot be developed because of land-use conflicts. Once a host rock has been chosen, the implementer considers the hydrogeologic environment and determines what, if any, engineered barriers are appropriate as well as how the repository system as a whole will be designed.[25] The implementer then is expected to advance its safety case, a set of arguments and analyses demonstrating why the proposed deep-minded geologic repository will isolate and contain HLW and SNF for as long as society demands.[26] (Various standards and regulatory requirements reflect those demands.) There is broad scientific agreement that deep-mined geologic repositories can be constructed in a wide variety of host-rock formations and hydrogeologic environments, including in salt, crystalline rock such as granite, different clay formations,

108    United States Nuclear Waste Technical Review Board

and unsaturated volcanic tuff.[27] A brief discussion of the work to evaluate each host-rock type for permanent disposal of HLW and SNF follows.

## Salt

Disposal of HLW and SNF in salt has been explored in detail in several countries for more than a one-half of a century. When the NAS first proposed developing such a repository, it noted that "... the great advantage is that no water can pass through salt. Fractures are self-sealing..." (NAS: 1957, 4-5; Appendix F). Those two properties have been at the core of the salt disposal concept adopted, for example, by the German waste-management program.[28]

### *Disposal Concept*

As originally articulated, the salt disposal concept appears to be extremely elegant in its simplicity. Put in the simplest terms, if the salt is there, water flow, the predominant mechanism for transporting radionuclides to the biosphere, is not occurring at rates of concern for waste disposal. It is then just a matter of carving out drifts in the formation. Waste is lowered and emplaced into the drifts, most likely in boreholes. The shafts leading to the repository, the drifts themselves, and the boreholes then are sealed with crushed and compacted salt.

Under lithostatic pressure exerted by the layers of rock above the repository, the salt flows slowly, closing around emplaced disposal packages and healing any fractures or voids that may have formed during the construction phase. Waste packages are not considered longterm barriers for isolating and containing HLW and SNF because localized brine inclusions could cause them to fail. Even so, because the environment in the repository would evolve over several hundred years from being oxidizing to become reducing, the waste would remain in a relatively insoluble form.

It is possible to analyze disposal of HLW and SNF in salt generically to determine whether it is an appropriate host rock for a deep-mined geologic repository.

Those generic studies would have to address questions about the effects of heat on brine migration and whether pressures become too great when hydrogen is generated as small amounts of water contact the waste. Moreover, extensive underground exploration of salt beds or domes would be required to determine whether a particular site might be suitable for a deep-mined geologic repository. But at least some participants in the German program believe that an

Experience Gained from Programs ...        109

undisturbed deep-mined geologic repository in salt will have zero release of radionuclides to the biosphere for at least one million years (Krone et al. 2008).[29]

### *Safety Case*

The safety case for disposing of HLW and SNF in salt involves a demonstration that the undisturbed geologic barrier would perform as anticipated and that engineered barriers (mainly shaft, drift, and borehole seals) would prevent brine inflow through the man-made penetrations of the salt barrier, thereby ensuring that any movement of dissolved radionuclides along those pathways will be minimal.

Sophisticated modeling work has been carried out to support the proposition that a disturbed salt repository holding non-heat-generating waste will isolate and contain radioactive waste for long periods of time (EPA 2006). Additional modeling appears to support the proposition for heat-generating waste as well, although that claim has not been subjected to a formal empirical or regulatory test.

Toward that end, experiments have been conducted to understand the compaction behavior of crushed salt. Investigations have been carried out to thermally simulate disposal in drifts of waste packages containing SNF. Based on these studies, experts from the German program argue that the engineered elements of the repository it contemplates using (other than the waste packages) would function reliably (Müller-Hoeppe et al. 2008). So far however, no country has advanced a comprehensive safety case for disposing of HLW and SNF in salt.

## Crystalline Rock

The KBS-3 plan developed by the Swedish program, supplemented by work carried out by the Canadian and Finnish programs, has resulted in a well-articulated crystalline rock disposal concept. The viability of this disposal concept depends on finding a site where the acidity and the oxidation-reduction potential of the groundwater enveloping the crystalline rock formation fall into an appropriate range.

In that case, according to the laws of thermodynamics, elemental copper would not react with the groundwater and thus waste packages made of that material would contain the SNF virtually forever. Sites that have few fractures add an extra layer of protection.

## Disposal Concept

KBS-3 envisions a repository system composed of multiple compatible natural and engineered barriers. Groundwater in Swedish crystalline basement rock possesses the requisite chemical and electrochemical properties. The crystalline rock, however, is not impenetrable. Fractures permit groundwater to flow within the typical formation, although the flow usually is quite slow, thereby limiting release of radionuclides to the environment.

To reduce further the release of radionuclides, engineered barriers are critical elements of the KBS-3 plan. Rings of bentonite clay are used to line the boreholes where the packages will be emplaced. This material further limits exposure of the copper canisters that contain the SNF to the groundwater. The bentonite also protects the canisters in the event of small movements in the rock and delays the spread of radionuclides that might escape from the waste package. The repository is designed so that the drifts can be backfilled with bentonite. The waste package features a canister, which is constructed from five-centimeter thick copper. In addition to being corrosion-resistant in the repository's environment, the canister can withstand some of the mechanical forces caused by the movement of the rock (Hedin 2008). Inside the copper canister is a nodular cast-iron insert to increase the mechanical strength of the waste package.

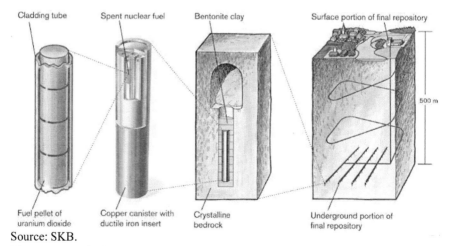

Source: SKB.
The KBS-3 method of disposing spent nuclear fuel.

## Safety Case

The safety case for the crystalline rock disposal concept depends most importantly on groundwater having favorable chemical properties, including

Experience Gained from Programs ... 111

acidity, oxidation-reduction potential, and dissolved solutes. All alternative corrosion mechanisms to the one articulated in the disposal concept have to be investigated to ensure that copper will remain in its elemental state. Bentonite has to be shown to limit advective transport under the chemical, thermal, mechanical, and hydrologic conditions expected to be present in the deep-mined geologic repository. Techniques for flawlessly welding lids on to the waste package must be demonstrated.

SKB has studied these issues. It has constructed laboratories to test the properties of bentonite under a wide range of conditions. It also has built a laboratory to investigate methods for welding and inspecting canister lids at an industrial scale. It has undertaken preliminary work to investigate the claim that new mechanisms have been identified by which copper might corrode in the basement rock. An extensive review of this particular issue, however, by the oversight body largely concluded that the claim was not technically supported (Swedish National Council for Nuclear Waste 2009). A report, however, to the Swedish Radiation Safety Authority (SSM) suggests that this controversy has not yet been put entirely to rest (Macdonald and Sharifi-Asi 2011).The KBS-3 plan has been subjected to rigorous national and international peer-review (see, for example, NEA 2001). SSM and the National Council for Nuclear Waste have regularly evaluated SKB's plans to address outstanding technical issues. Neither organization has found flaws with the disposal concept that would require its abandonment or radical revision (see, for example, Swedish National Council for Nuclear Waste 2008). Confidence in the concept is increased because intact copper nodules, millions of years old, have been found enclosed in the same type of formations that might someday host a deep-mined geologic repository.

All information obtained through its research program was analyzed by SKB as it prepared a license application, which was submitted to the authorities in March 2011. As part of that application, SKB carried out a quantitative postclosure safety analysis of the proposed facility, primarily by estimating the possible dispersion of radionuclides and how those releases would be distributed in time for a representative selection of future potential scenario sequences.

## Clay

A repository mined out of a layer of clay or clay-like materials may be an effective approach to isolating and containing HLW and SNF because the rock

# 112      United States Nuclear Waste Technical Review Board

has three important properties. First, water typically moves very slowly through clay. Second, clay can have a high sorption capacity for radionuclides. Third, any fissures or fracture planes in the rock close by themselves over the course of time. Three countries are actively investigating the possibility of developing a deep-mined geologic repository in clay formations found within their borders: Belgium (boom clay); France (argillite); and Switzerland (opalinus clay). Although some important differences exist among the three countries' disposal concepts and the types of waste they will dispose of, the similarities in the disposal concepts are more substantial. For the purposes of the discussion below, Switzerland's national program is described (NAGRA 2002).

### *Disposal Concept*

The absence of significant advective groundwater flow in the clay means that radionuclides would move out of the engineered barriers and the undisturbed rock at a very slow rate.

Any such movement would thus be controlled by diffusion, which suggests that only the most mobile and longest-lived radionuclides can reach the edge of the clay formation. Rock units surrounding the formation where a repository might be built, which also are rich in clay, would further slow the release of radionuclides to the biosphere. Chemical conditions in clay would be reducing, thereby maintaining the constituents of SNF in a low-solubility state.

Engineered barriers would be designed to work with the natural ones. The waste package would be constructed from steel and expected to prevent the inflow of water for several thousands of years.

When the packages start to corrode, the resulting corrosion products might hydrolyze to create more acidic and aggressive near-field environments. The packages would be emplaced horizontally, and the drifts would be backfilled with bentonite. As with the crystalline rock disposal concept, the bentonite would retard the radionuclides and ensure that their transport is only by diffusion.

A separate but co-located, pilot facility would be constructed after a site for a deep-mined geologic repository is selected. Representative volumes of waste would be disposed of in this facility.

Monitoring would take place to validate long-term predictions of how the host rock is evolving as well as to identify possible early indications of safety barrier failures.

## Safety Case

Like the German waste-management program, the Swiss program is intended to demonstrate that there will be zero releases for at least a million years if the host rock is undisturbed (NAGRA 2002). Only if there is significant climate change, borehole penetration of the repository, or deep groundwater extraction at the site would there be any release to the biosphere. Work to enhance the technical basis for the clay safety case continues, especially for repository performance under disturbed conditions. The following are some of the key uncertainties being explored:

- Solubility limits and sorption coefficients
- Rate at which the clay is resaturated
- Impact of heat on the performance of the bentonite buffer
- Gas generation by steel canister corrosion.

In creating its safety case, the waste-management program in Switzerland explicitly develops multiple lines of argument, including the use of alternative indicators that are complementary to those of dose and risk; natural analogues; and conservative performance assessments. Preliminary safety assessments using both deterministic and probabilistic methodologies have been carried out. Based on those assessments, the Swiss government has accepted the safety case advanced by NAGRA. The safety case also has been peer reviewed by an international team assembled by the NEA (NEA 2004c). The safety case also forms the basis for the Sectoral Plan, which currently guides efforts to identify suitable sites (SFOE 2008).

## Unsaturated Volcanic Tuff

The United States is the only country that has developed a safety case for disposing of HLW and SNF in a deep-mined geologic repository located in unsaturated volcanic tuff. That safety case was developed by DOE in parallel with the characterization of a specific site located at Yucca Mountain in Nevada.

The safety case rests on two main pillars. First, engineered barriers minimize the amount of water that can come in contact with the HLW and SNF. Second, transport of radionuclides to the biosphere is limited by the amount of water leaving the drifts (Abraham 2002).

Source: DOE.
View of Yucca Mountain from Busted Butte.

## *Disposal Concept*

Very little precipitation falls on Yucca Mountain. A large fraction of what does returns to the atmosphere by evaporation, plant transpiration, and run-off; only a small amount of water infiltrates below the root zone, and even less seeps into the repository drifts. The location of the proposed repository lies in the unsaturated zone, where the environment is oxidizing, so the constituents of SNF would react with oxygen and become more mobile. To limit the release of radionuclides, corrosion-resistant titanium drip shields would be installed to divert the water that enters the drifts, thereby protecting the waste packages that lie underneath. The packages themselves would be fabricated with an outer layer of a nickel-based material, Alloy 22, and an inner layer of stainless steel. The packages would degrade very slowly in repository environments because of this corrosion-resistant alloy. Any radionuclides that escaped would move slowly because significant advective transport is unlikely.

## *Safety Case*

Relying on this disposal concept, DOE issued a final environmental impact statement (DOE 2002). That assessment was one of the reasons that Congress approved the selection of the Yucca Mountain site (U.S. Congress

2002). In June 2008, DOE submitted a license application to NRC based on a Total System Performance Assessment, which is grounded on more than 30 years of site-specific scientific and technical investigations.

Nonetheless, questions remain about the safety case for unsaturated volcanic tuff. Nearly 300 issues have been raised by supporters and opponents participating in the licensing hearing convened by the NRC. Moreover, the NWTRB, which is not a party to those hearings, has noted that there is only a poor understanding of how fast water moves in the unsaturated zone (NWTRB 2008, 30-31). It also has suggested that deliquescence-induced localized corrosion could lead to more-rapid waste package degradation than DOE maintains (NWTRB 2008, 25-28). These issues might eventually be resolved in the course of that hearing process. For the moment, at least, that hearing process has been suspended.

Table 2 presents a summary of the information contained in this section.

### Table 2.Safety cases for disposing of HLW and snf in a deep-mined geologic repository

| HOST ROCK | COUNTRIES ADOPTING SAFETY CASE | FOUNDATIONS OF SAFETY CASE |
|---|---|---|
| Salt | Germany | Presence of salt implies the absence of flowing water Fractures are self-sealing High thermal conductivity and high acceptable design temperatures permit the construction of a small-footprint repository |
| Crystalline rock | Sweden, Finland | Near-neutral, reducing groundwater thermodynamically precludes corrosion of elemental copper Bentonite clay limits water contact with the waste package and retards the release of any radionuclides |
| Clay | Belgium, France, Switzerland | Water typically moves very slowly through clays Clay can have a high sorption capacity for radionuclides Any fissures or fracture planes present in the rock close by themselves over time |
| | | Simplified design has no or small requirements for engineered barriers |
| Unsaturated volcanic tuff | United States | Under current climatic conditions, limited amounts of rain fall on Yucca Mountain |
| | | Engineered barriers prevent the release of radionuclides to the natural system for a long time and release radionuclides slowly after penetration Transport of radionuclides to the biosphere is limited by the amount of water leaving the drifts |

# Major Summary Points

- Disposal concepts currently under consideration vary depending on the host rock within which the repository would be constructed.
- In all national programs, the disposal concept relies on both natural and engineered barriers. Concepts differ markedly in how much each barrier contributes to waste isolation and containment.
- Some disposal concepts, such as the KBS-3 approach to developing a repository in crystalline rock, are supported directly by fundamental physical principles. Others are supported by detailed modeling and analyses.

# SUBSTANCE AND ADOPTION OF HEALTH AND SAFETY STANDARDS AND REGULATIONS

Health and safety regulations serve two purposes. They record society's views about what constitutes acceptable risk, and they establish mechanisms for certifying that an implementer's plan to develop a deep-mined geologic repository can, with a high degree of confidence, satisfy those requirements. All of the 13 national waste-management programs have put in place at least a rudimentary regulatory regime.

## Substance of the Regulations

As nations adopted geologic disposal as their preferred option for the long-term management of HLW and SNF, the question soon arose about what standards should govern that choice. Many of those involved argued that it is not the responsibility of the nuclear community to "purify" the earth by reducing the population's total exposure to radioactive materials. Thus they promoted the position that a repository need not sequester waste beyond the time when it becomes less hazardous than the uranium ore from which it was derived (see, for example, the discussion in NAS 1983). This simple approach has been replaced by more-sophisticated standards that focus on four specific issues.[30]

First, the regulators must decide on the length of the compliance period, the time over which the repository is expected to satisfy the protective standards. Regulators originally chose compliance periods of several thousand years. Now national programs in Belgium, France, Germany, Sweden, Switzerland, and the United States have selected compliance periods of at least 100,000 years and, in most of those cases, as much as 1,000,000 years. In the United Kingdom, the regulator requires the implementer to choose a compliance period and justify its choice.

Second, regulators impose dose constraints or risk limits or some combination of the two. Increasingly national waste-management programs have converged on similar dose constraints and risk limits, at least for the first few thousands of years that a repository is expected to isolate and contain HLW and SNF. Dose constraints vary between 0.1 and 0.3 millisieverts per year. Risk limits vary between a probability of $10^{-5}$ and $10^{-6}$ per year that death or serious health effects will arise over the course of the lifetime of an individual from exposure to radionuclides released from a repository. Beyond the first few thousand years, dose constraints may be also much as 1 millisievert per year and risk limits may be as low as $10^{-3}$ per year for some scenarios.

Third, the regulators decide how prescriptive their requirements should be. Those choices have produced considerable variation in how much direction the regulators provide. In some national programs, most notably the one in the United States, the rules are quite detailed, laying out specific requirements that the implementer must fulfill in order to get permission to construct or operate a deep-mined geologic repository. Among the specifics called for are the following (NRC 2001).

- Information that must be contained in an application to construct a repository
- Technical criteria for conducting preclosure and postclosure performance analyses, testing designs, and carrying out monitoring
- Requirements for evaluating the consequences of inadvertent human intrusion into a repository
- Types of institutional controls over the repository.

One of the ways that regulators defend the imposition of prescriptive requirements is by maintaining that they reduce the implementer's uncertainty.

In other national programs, such as the ones in Canada and Germany, the regulators simply require the implementer to provide information and to perform analyses in whatever manner it chooses. For example, in Germany, the implementer is told that "... a comprehensive safety case [must] be

118 United States Nuclear Waste Technical Review Board

documented for all operating states of the final repository..." and that it must provide "... analysis and representation of the robustness of the final repository system... [along] with the reasons why this assessment is to be trusted..." (BMU 2010, 11-12). One of the ways that regulators defend the imposition of non-prescriptive requirements is by maintaining that they increase the implementer's flexibility.

Fourth, national programs differ in how compliance with the standards is to be demonstrated and what the requirements for demonstration are. In the Finnish, French, and German programs, a small set of scenarios is defined, and the implementer is required to evaluate deterministically how its proposed deep-mined geologic repository would perform. The regulators in the United States believe that the appropriate compliance methodology is a probabilistic performance assessment. Finally, the regulators in some countries, including Sweden and France, call for mixed approaches in which quantitative methods are used to evaluate performance early in the compliance period and more-qualitative methods are applied to later time periods. The Canadian regulators leave the choice of methodology completely up to the implementer.

## Adoption of Regulations

All national programs that have adopted specific regulations for the long-term management of HLW and SNF have done so in similar ways. Traditional mechanisms for public involvement are used. Most typically, the regulators themselves prepare a draft of the rules they want to adopt. Rarely will other interested and affected parties have the right to comment while the draft is being prepared.[31] Once it is completed, the regulators generally will release the proposed requirements for public comment. In some cases, the regulators convene open meetings where parties can ask the regulators questions and provide recommendations directly.

Regulators review the public input and determine what changes, if any, need to be made to the draft rules. Sometimes they also will prepare a document that explains why some suggestions were accepted and others were not. Once a final rule is published, it will take effect after a short period passes. In some countries, parties that disagree with the regulators' determinations can appeal to the courts. In the United States, this tactic tends to be used more often than not. Although courts there may overturn or remand regulations to their authors, the judiciary usually defers to the regulators' expert judgment as long as there is a reasoned basis for it.

National programs typically have put health and safety regulations into place *after* the implementer has begun to formulate its safety case or to identify candidate sites for a deep-mined geologic repository but *before* specific sites have been chosen. Some interested and affected parties may contend that revising regulations *after* a final site has been selected is inappropriate if the change process is not well explained and supported.

What transpired in the United States provides an instructive example of this dynamic. In response to Congress' passage of the 1992 Energy Policy Act, the Environmental Protection Agency and NRC issued Yucca Mountain-specific standards and regulations in the mid-2000s. Both agencies maintained that the new rules reflected the state-of-the art.[32] The State of Nevada, an opponent of the repository, claimed that the switch in approach was deliberately designed to aid in the licensing of the facility. In particular, the State argued that new technical information had come to light that, under one provision in the old—but not the new—rules, would have required the Secretary of Energy to disqualify the site (Miller 1996).[33] For some, this argument was made more persuasive because of DOE's deep involvement in the interagency process that reviewed the rules as the regulators were developing them (EPA 2005). Of course, sorting out motivations is never easy. For that reason, trust and confidence in the regulators may be compromised for at least some interested and affected parties if the process for changing rules is not completely transparent.

## Major Summary Points

- National programs have adopted comparable health and safety standards and regulations. Some differences, however, exist both in the level of acceptable dose or risk and in the compliance period. The length of the compliance period typically has become much longer as the rules have evolved.
- A major difference between the regulatory process in the United States compared to most other countries is the requirement that a quantitative, probabilistic assessment of compliance with the standard must be presented. Compared to most other nations, the regulatory regime in the United States is very prescriptive.

120 United States Nuclear Waste Technical Review Board

The process and rationale used to modify requirements can affect the public's trust and confidence in the regulator, especially once a site has been chosen for a deep-mined geologic repository.

## STRATEGIES FOR IDENTIFYING CANDIDATE SITES FOR A DEEP-MINED GEOLOGIC REPOSITORY

President Jimmy Carter's Interagency Review Group on Nuclear Waste Management (IRG) observed that site-selection strategies for a deep-mined geologic repository necessarily involve passing candidates through what is, in effect, two distinctly different "filters." On the one hand, detailed and quantitative technical requirements have to be met. On the other, sites could be disqualified because of considerations such as the "... lack of social acceptance, high population density, or difficulty of access." The two filters could be applied in any order. In the IRG's view, at least, although the suite of sites eventually selected might be different, depending on the order in which the filters were applied, "... equally suitable sites should emerge from either approach..." (IRG 1979 80; 81).[34] Over the years, the United States and other nations have initiated roughly two-dozen efforts to identify or to create processes for identifying potential repository sites. What is noteworthy is how varied those efforts have been.

### Technical Filter

Part of the variation stems from how the technical filter is constructed. In some cases, efforts to identify candidate sites have focused from the beginning on specific host-rock formations. The choice of those formations has been dictated either by constraints imposed by a country's geology or land-use patterns, by a view that particular host-rock formations possess distinctive advantages in terms of isolating and containing HLW and SNF, or by a combination of these rationales. In Sweden and Finland, efforts were concentrated exclusively on the granitic formations of the Baltic Shield, which underlies the vast majority of both countries (Sundqvist 2002). In their early attempts to identify a candidate site, the United States and Germany looked only at locations in salt formations. The United States investigated sites near Lyons, Kansas, and in the Permian Basin in the southwest. The Germans chose a site at

Gorleben. Because of the political turbulence surrounding the development of nuclear power, however, investigations at that site were suspended for 10 years. Those studies have been resumed recently (BfS 2010). The Swiss initially looked at sites in both crystalline rock and clay. Now only clay sites are being considered (SFOE 2008).

In other cases, efforts to identify candidate sites cast the net more broadly by enumerating generic qualifying and disqualifying conditions.[35] Qualifying conditions must be satisfied for a candidate site to be considered acceptable; disqualifying conditions eliminate a candidate sites from further consideration. In France, qualifying conditions included the depth of the repository horizon, the thickness of the host formation, the absence of natural resources, and present and future hydrogeologic flow patterns. In the United States, before the passage in 1987 of the Nuclear Waste Policy Amendments Act (NWPAA), both qualifying and disqualifying conditions for the preliminary screening of potential sites were elaborated in considerable detail (DOE 1984). The Canadians have advanced six "safety factors" that are likely to be used to evaluate candidate sites (NWMO 2009). In the Japanese screening approach, sites within a 15 km radius of the center of Quaternary volcanoes are rejected (NUMO 2004). Exclusionary screening criteria eliminate locations in England and Wales that are close to natural resources, such as fossil fuels and fresh water, as well as those within deep karstic formations and known source rocks for thermal springs (UKG 2008). In Germany, the AkEnd working group proposed seven qualifying conditions, which then were used to eliminate from consideration host-rock formations in five regions (AkEnd 2002).[36] Table 3 summarizes how the technical filter has been constructed in the two-dozen siting initiatives undertaken to date.

## Nontechnical Filter

Just as the construction of the technical filter introduces considerable variation in strategies for selecting candidate sites for a deep geologic repository, so does the construction of the nontechnical one. Arguably this filter's most important property relates to the power that a state or a community can exercise. In particular, the experience in the United States illustrates that the power need not be formalized in order to be effective. In 1971, the AEC proposed a site near Lyons, Kansas. Neither the state nor the local community—both of which opposed the selection—was given any explicit power to object.

Source: SKB.
Determining the suitability of granite as a host rock.

Nonetheless, opposition by the state contributed to a decision not to develop the site. Two years later, the Energy Research and Development Administration (ERDA) sought to investigate sites in the Salina Basin, which runs through Michigan, New York, and Ohio. Although the governors of those states did not have any formal authority, their opposition was sufficient to prevent any site investigations (Carter 1978).

Based on this experience, the IRG advanced the idea of "consultation and concurrence" in its recommendations to President Carter. Although this concept was never interpreted by the administration as offering a state the formal power to reject a site, as a practical matter, it did precisely that. In the IRG's draft report to the President, it observed that "...a state would be in agreement with each step of the [repository development] process before the next activity..." would begin (IRG 1979, 88). Later, "consultation and concurrence" evolved into the less potent concept of "consultation and cooperation," a formulation that was adopted in the 1982 NWPA.

That legislation gave a state formal veto power over the selection of a site by the President, but the state veto could be overridden by a resolution approved by both houses of Congress. In the NWPAA, Congress mandated that only the Yucca Mountain site could be characterized, although it left unchanged Nevada's power to veto the President's recommendation, subject to an override by Congress.

## Table 3. Technical filter

| Country | Focus on hostrock formation | Qualifying and disqualifying conditions formally adopted To structure The site-selection process |
|---|---|---|
| United States | | |
| AEC: Lyons | Salt | None |
| ERDA: Permian Basin | Salt | None |
| IRG | | Recommended that NRC establish siting criteria. |
| DOE: NWPA | | DOE established detailed siting criteria. |
| DOE: WIPP | Salt | None |
| DOE: NWPAA | Tuff | DOE established new siting criteria. |
| Belgium | Clay | None |
| Canada | | |
| AECL | Crystalline rock | None |
| NWMO | | Proposed general siting criteria.* |
| China | Crystalline rock | National Nuclear Safety Administration established general siting criteria. |
| Finland | Crystalline rock | Radiation and Nuclear Safety Authority (STUK) established general siting criteria. |
| France | | |
| CEA | | No siting criteria established. |
| ANDRA | Crystalline rock and clay | Nuclear Safety Authority (ASN) established general criteria. |
| Germany | | |
| Gorleben (1970s) | Salt | None |
| AkEnd | | Proposed specific siting criteria. |
| Gorleben (2010) | Salt | None |
| Japan | | NUMO proposed general siting criteria. |
| Republic of Korea | No decision made. | No decision made. |
| Spain | No decision made. | No decision made. |
| Sweden | | |
| SKB: Pre-1992 | Crystalline rock | None |
| SKB: Post-1992 | Crystalline rock | None |
| Switzerland | | |
| SFOE: Pre-2005 | Crystalline rock and clay | None |
| SFOE: Post-2005 | Clay | General siting criteria established. |
| United Kingdom | | None |
| UKAEA: 1970s | | |

# 124    United States Nuclear Waste Technical Review Board

**Table 3. (Continued)**

| Country | Focus on hostrock formation | Qualifying and disqualifying conditions formally adopted To structure The site-selection process |
|---|---|---|
| NIREX: Late 1980s | Wide variety of sedimentary and igneous rocks | None |
| NDA | Wide variety of sedimentary and igneous rocks | Siting criteria advanced as Government policy. |

*The Canadian Nuclear Safety Commission has issued guidance on siting a deep-mined geologic repository for HLW and SNF.

Since the early 1990s, nations other than the United States increasingly have constructed their nontechnical filters in ways that significantly empower local jurisdictions. In Sweden, in the 1980s, SKB unilaterally identified potential sites and attempted to drill boreholes to test whether the sites are suitable for a deep-mined geologic repository. Subsequently, however, the implementer reconciled itself to the fact that the local communities have formal veto authority over the siting of infrastructure projects. Consequently, in the early 1990s, SKB issued a call for volunteers to host a repository. A number of communities responded positively, knowing that they held veto power over the final selection of a site. (Although the veto can be overridden by the national government, both political tradition and the legal constraints imposed make this an unlikely event.) A similar situation exists in Finland.

In France, districts and communities near Bure volunteered to host an underground research laboratory, knowing that it might become the forerunner to a repository. In Switzerland, now that technically suitable regions have been identified, cantonal commissions are being established to participate in decisions leading to the selection of at least two candidate sites. Although the presumption is that efforts will be made to resolve any conflicts that may arise, the final decision will rest with the Federal Council, subject to a possible national referendum.[37] Newly adopted processes for identifying candidate sites in England and Wales in the United Kingdom, Japan, and Canada all require localities to volunteer even before paper studies of the local geology can be conducted.

Experience both in the United States and in other nations suggests that communities already hosting nuclear facilities or communities where benefits might make a significant economic or social difference may be especially receptive to being considered as a candidate repository site.[38] In Sweden and Finland, candidate sites were identified in communities where nuclear power plants operate. In Britain, borough and county councils in West Cumbria, where the Sellafield nuclear facilities are located, expressed an early interest in being considered a potential repository site. The state of Lower Saxony in Germany welcomed investigations of the Gorleben site. In the United States, the city of Carlsbad, New Mexico, aggressively sought to be considered as the location for WIPP as a means of bringing employment to the community.[39]

Focusing primarily on the nontechnical filter does not necessarily eliminate the frictions associated with identifying candidate sites for a deep geologic repository.[40] In Japan, the situation is very much in limbo. In 2002, Government launched a voluntary process. The mayor of the Toyo township in the Kochi Prefecture southwest of Tokyo announced that he would respond positively to NUMO's open solicitation. Opposition arose immediately within the local community and from governors of nearby prefectures. Ultimately the mayor was soundly defeated in an election that served as a referendum on participation in the site-selection process. No other community has stepped forward since. The Japanese government is now suggesting that alternatives to a voluntary approach may have to be considered. In the United States, although two of the three candidate sites identified for the first repository under the NWPA were "nuclear communities," governors of the two states where they were located announced their intense opposition, and both made clear that they would exercise a veto if the site in their site were chosen for development.[41]

Table 4 summarizes how the nontechnical filter has been constructed in the two-dozen siting initiatives.

## Interdependence of the Filters

The two filters are not independent of each other, except in some abstract or theoretical sense. The construction of the nontechnical one may affect the technical one in important ways. To begin with, applying the technical and nontechnical filters is not purely mechanical nor can it typically be programmed neutrally. In the United States, to consider just one example, the technical content of the 1984 siting guidelines (10 CFR 960) was constrained by language in the NWPA and was profoundly influenced by an administrative rulemaking process, in which states sought to include guidelines that would lead to the disqualification of sites within their borders.[42] In addition, using those guidelines to narrow the nine potential sites to five sites, which would be evaluated, to three sites, which would be characterized, revealed how difficult it

126 United States Nuclear Waste Technical Review Board

can be to implement even an ostensibly objective and technical process (DOE 1986a; Merkhofer and Keeney 1987). The evaluation of the qualifying and disqualifying conditions was open, perhaps unavoidably, to considerable interpretive flexibility. Moreover, by focusing on selective and discrete attributes of a candidate site, the performance of the total repository system never received appropriate attention.

Further, implicit in a voluntarist approach is the presumption that a very wide range of geologic features and locations are suitable or can be made suitable. In some cases, this presumption is well-founded. In both Sweden and Finland, the Baltic Shield is so pervasive that sites throughout both countries likely could be found to develop a repository. Similarly, in the United States, the salt formations of the Salina and Permian Basins are so extensive that many locations could be considered. In other cases, even after taking into account fairly general disqualifying conditions, potential disconnects may very well arise, so that applying *both* the technical and the nontechnical filters yields a null set of potentially suitable and acceptable sites. For many countries, including the United States under the NWPA and Switzerland, applying the technical filter rigorously at the start is viewed as a prerequisite for creating a credible process for identifying candidate sites for a deep geologic repository. Yet, in the former instance, Congress halted characterization of any site other than Yucca Mountain. In the latter one, the final outcome is still to be determined.

**Table 4. Nontechnical filter**

| Nontechnical Filter | | | | |
|---|---|---|---|---|
| country | Early state or local involvement in The process | Volunteer | Nuclear community | Benefits package |
| United States | | | | |
| AEC: Lyons | None | No | No | None |
| ERDA: Permian Basin | None | No | No | None |
| IRG | Non-site specific process. | Proposed consultation and concurrence. | Not applicable | Not applicable. |
| DOE: NWPA | None | No | Two nuclear communities were | Benefits package |

| Nontechnical Filter | | | | |
|---|---|---|---|---|
| country | Early state or local involvement in The process | Volunteer | Nuclear community | Benefits package |
| | | | among the three final sites selected. | could be negotiated. |
| DOE: WIPP | Informal | Carlsbad community expressed willingness to host facility. New Mexico did not object. | No | Benefits package received. |
| DOE: NWPAA | None | No | Yes | Benefits package could be negotiated. |
| Belgium | No decision made. | No decision made. | No decision made. | No decision made. |
| Canada | | | | |
| AECL | Non-site specific process. | Not applicable. | Not applicable. | Not applicable. |
| NWMO | Extensive | Yes | No decision made. | No decision made. |
| China | No decision made. | No decision made. | No decision made. | No decision made. |
| Finland | Yes | Did not veto selection. | | |
| France | | | | |
| CEA | No | No | No | No |
| ANDRA | Yes | Volunteers were sought to host an underground research laboratory. | No | Benefits package was received. |
| Germany | | | | |
| Gorleben (1970s) | Yes. | Yes | No | Benefits package received. |
| AkEnd | Non-site specific process. | Not applicable. | Not applicable. | Not applicable. |
| Gorleben (2010) | Yes | Yes | No | Benefits package is available. |
| Japan | Yes | Volunteer unsuccessfully sought for the last eight years | No decision made. | Benefits package is available. |
| Republic of Korea | No decision made. | No decision made. | No decision made. | No decision made. |

# 128 United States Nuclear Waste Technical Review Board

## Table 4. (Continued)

| Nontechnical Filter | | | | |
|---|---|---|---|---|
| country | Early state or local involvement in The process | Volunteer | Nuclear community | Benefits package |
| Spain | No decision made | No decision made | No decision made | No decision made |
| Sweden | | | | |
| SKB: Pre-1992 | No | No | No | No |
| SKB: Post-1992 | Yes | Originally eight communities volunteered for feasibility studies to evaluate whether they would be prepared to host a repository. Two of the communities volunteered to host a repository. | Yes | Benefits package was negotiated in which the community not selected to host a repository would receivethree-quarters of the total. |
| Switzerland | | | | |
| SFOE: Pre-2005 | No | No | No | No |
| SFOE: Post-20 5 | Communities will be con ulted as part of Phase I of the Sectoral Plan | | No decision made. | No decision made. |
| United Kingdom | | | | |
| UKAEA: 1970s | No | No | No decision made. | No |
| NIREX: Late 1980s | No | No | Yes | No |
| NDA | Yes | Borough and county councils located near the Sellafield site have | Yes | Benefits package is available. |

| Nontechnical Filter | | | | |
|---|---|---|---|---|
| country | Early state or local involvement in The process | Volunteer | Nuclear community | Benefits package |
| | | expressed an interest in possibly hosting a repository. | | |

## Sequencing

An additional source of variation among national programs can be traced to policies that govern the sequence for accepting or rejecting a candidate site. In the United States, a "serial" policy was at first adopted. Sites would be evaluated formally one by one until a suitable site was found. The Carter Administration, however, proposed a "parallel" approach, in which at least two candidate sites would be characterized simultaneously and compared. This recommendation was ratified in the NWPA. Several years later, Congress returned to a serial policy with the passage of the NWPAA, which singled out Yucca Mountain in Nevada as the sole site to be characterized. Although several sites were informally considered in Germany during the 1960s and 1970s, the Gorleben site was the only one investigated seriously. Despite a clear call for communities to volunteer, the government in the United Kingdom has received an expression of interest from only one group of local governments. Thus, the opportunity to compare candidate sites may never materialize. Conversely, the Swiss mandate the parallel characterization and comparison of at least two sites. In Sweden and Finland, sites in two and four municipalities respectively were investigated in comparable detail.[43] In France, more than 30 sites were studied before 1990. Legislation passed in 1991 required the comparison of a clay site and a crystalline rock site. Finding an appropriate site in crystalline rock was not possible, so an actual clay site was ultimately contrasted with a hypothetical crystalline rock site. Finally, in several countries, including Canada, Belgium, China, and Britain, no explicit decision has been made about using a serial or a parallel approach.

Table 5 summarizes how the filters have been or are being applied by the 13 nations considered in this report.

# Table 5. Processes for selecting candidate sites for a deep-mined geologic repository

| COUNTRY | APPLICATION OF THE FILTERS | CHALLENGES ENCOUNTERED | OUTCOME | SERIAL OR PARALLEL EVALUATION OF SITES |
|---|---|---|---|---|
| United States | | | | |
| AEC: Lyons | Technical filter was applied. | State opposition led to Congress to deny appropriations. | Site was abandoned. | Serial |
| ERDA: Permian Basin | Technical filter was applied. | State opposition arose. | Sites were never investigated. | Serial |
| IRG | Asserted that order in which the filters were applied did not matter. | Proposals were generally accepted. | Most proposals were incorporated into the NWPA. | A parallel approach was proposed. |
| DOE: NWPA | Technical filter was applied. | Costs of site characterization rose; political opposition from final three candidate states intensified. | Congress amended the NWPA to allow only the Yucca Mountain site to be characterized. | Parallel |
| DOE: WIPP | Nontechnical filter was applied once the AEC had identified salt as a favorable generic host-rock formation. | Local community supported the selection of the site, overcoming opposition from the Carter Administration. | WIPP began accepting waste from the DOE defense complex in 1999. | Serial |
| DOE: NWPAA | Nontechnical filter was applied. | State of Nevada opposed the selection of Yucca Mountain. | Administration moved to withdraw license application in 2010. Litigation is ongoing at the NRC and in the courts. | Serial |
| Belgium | No decision made. | Not applicable. | Not applicable. | No decision made. |
| Canada | | | | |
| AECL | Technical filter was applied first. | Disposal concept was deemed socially unacceptable. | National waste-management program was reconstituted. | No decision made. |
| NWMO | General siting criteria are used to identify areas that | Adaptive Phased Management approach generally accepted. | Siting process has been initiated. | Depends on the number of volunteers. |

| COUNTRY | APPLICATION OF THE FILTERS | CHALLENGES ENCOUNTERED | OUTCOME | SERIAL OR PARALLEL EVALUATION OF SITES |
|---|---|---|---|---|
| | might be suitable; eight volunteer communities have expressed an interest in learning more about the possibility for hosting a repository. | | | |
| China | Technical filter was applied. | No challenges have emerged to date. | Process is continuing. | No decision made. |
| Finland | Technical filter was applied; negotiations with communities took place. | Few challenges were encountered to date. | Site at Eurajoki is being developed for a repository. | Parallel |
| France | | | | |
| CEA | Technical filter was applied first. | Political opposition arose. | Sites were never investigated. | Serial |
| ANDRA | Nontechnical filter was applied. | Few challenges were encountered to date. | Underground research laboratory established in Meuse/Haute-Marne region; nearby a repository is being developed. | Parallel in theory; serial in practice. |
| Germany | | | | |
| Gorleben (1970s) | Technical filter was applied to select salt as the generic host rock. State of Lower Saxony proposed the specific site. | Although the local state was initially supportive, difficulties arose at the national level following the 1998 federal election. | Site investigations were suspended for 10 years. | Serial |
| AkEnd | Proposed a set of site-selection criteria to winnow down the number of potential sites; negotiations with communities would follow. | Political support for proposal never developed. | Proposal was never adopted as national policy. | Parallel |

Table 5. (Continued)

| COUNTRY | APPLICATION OF THE FILTERS | CHALLENGES ENCOUNTERED | OUTCOME | SERIAL OR PARALLEL EVALUATION OF SITES |
|---|---|---|---|---|
| Gorleben (2010) | Application of technical filter was resumed in October 2010. | Political opposition has arisen, but it is too early to know how intense and sustained it will be. Both the Federal and Lower Saxony governments support the construction of a deep-mined geologic repository at Gorleben. | Unclear | Serial |
| Republic of Korea | No decision made. | Not applicable | Not applicable | No decision made. |
| Spain | No decision made. | Not applicable | Not applicable | No decision made. |
| Sweden | | | | |
| SKB: Pre-1992 | Technical filter applied. | Political opposition arose. | Sites were never investigated. | No decision made. |
| SKB: Post-1992 | Nationwide search held for a community that would allow site investigations. | Many communities volunteered in the initial round. Two communities remained in contention until the end of the process. | Site in Östhammer is being developed for a repository. | Parallel |
| Switzerland | | | | |
| SFOE: Pre-2005 | Technical filter applied. | Criticism arose that there was no comparison of sites. | Government accepted the disposal concept but required site comparisons. | Serial |
| SFOE: Post-2005 | Technical filter being applied to be followed by the application of the nontechnical filter. | Few challenges have been encountered to date. | Unclear | Parallel in theory. |

| COUNTRY | APPLICATION OF THE FILTERS | CHALLENGES ENCOUNTERED | OUTCOME | SERIAL OR PARALLEL EVALUATION OF SITES |
|---|---|---|---|---|
| United Kingdom | | | | |
| UKAEA: 1970s | Technical filter to be applied first. | Planning permission was difficult to obtain. | | |
| Nirex: Late 1980s | Technical filter to be applied first. | Planning permission was denied. | | |
| NDA | General siting criteria will be used to identify areas that might be suitable. | No overt opposition to process. | Two adjacent communities are discussing with Government the possibility of participating formally in the Managing Radioactive Waste Safely process. | Depends on the number of volunteers communities. |

## Major Summary Points

- National programs must balance specificity and generality in constructing technical filters. More-specific qualifying and disqualifying conditions reduce the number of potential candidate sites but provide explicit guidance on a technical basis for their selection. General qualifying and disqualifying conditions allow more sites to be considered but typically provide only a generic technical basis for their selection.
- Many national programs seek to identify a suite of candidate sites so that they can be compared. This objective has not always been met. No country to date has fully characterized at depth more than one site

Although the process for identifying candidate sites appears clear and accountable, in practice it has proven to be cumbersome. The overwhelming majority of efforts have not succeeded. In many cases, the underlying cause of failure has been an inability to integrate the application of the technical and nontechnical filters, the most important of which is public support or opposition.

## SITE SELECTION FOR A DEEP-MINED GEOLOGIC REPOSITORY

In all national programs, the implementer is responsible for proposing a site for development as a deep-mined geologic repository. In some cases, political ratification at the national level of that decision also must take place.

## The Implementer's Decision

If only one site has been fully characterized at depth, as is the case in the French and American programs, it will be proposed by default if the implementer believes it to be suitable. Canada, Japan, and the United Kingdom, which have adopted, at least in principle, a parallel approach to identifying candidate sites, have not specified how the implementer will choose among multiple suitable sites. The implementer in Switzerland has been given only the most general instructions for making its decision (SFOE 2008, 64).

[The implementer shall] select the site for repository construction from the sites that have been integrated into the *Sectoral Plan* as an interim result. To be able to make and justify this selection, the level of knowledge for the different sites has to sufficient to allow a comparison to be carried out... The results—together with the evaluation of further aspects in accordance with the conceptual part of the *Sectoral Plan*—lead to an overall evaluation for site selection by the [implementer].

Implementers in Finland and Sweden, however, did decide among several candidate sites. Moreover, in the United States, before Congress only permitted the characterization of Yucca Mountain, DOE narrowed candidate sites from nine to five to three.

### *Finland*

Posiva Oy used the environmental impact assessment process to document its choice of Olkiluoto site in Eurajoki over Romuvaara in Kuhmo, Hästholmen in Loviisa, and Kivetty in Äänekoski (Posiva Oy 1999). The assessment considered the following eight criteria.

- Long-term safety
- Constructability
- Possibilities to expand the repository
- Operation of the final disposal facility
- Social acceptance
- Land use and environmental loading
- Infrastructure
- Costs.

Although the four sites have different geologic properties, they did not differ appreciably in terms of their long-term ability to isolate and contain SNF. Nor did the sites differ substantially in the geotechnical work needed to build the facility, the ease in which it could be expanded, infrastructure requirements, or costs. Because Eurajoki and Loviisa host nuclear power plants, their residents were less fearful of the transport of SNF through their communities and were more generally accepting of a repository. Developing a deep-mined geologic repository in Kuhmo and Äänekoski would significantly transform heavily forested areas. In the final analysis, Posiva chose the

Eurajoki site over the Lovissa one based on economic and development considerations rather than for technical reasons.

Source: SKB.
SKB informs the mayors from two candidate municipalities of its site-selection decision.

## *Sweden*

In 2009, SKB compared a site at Forsmark in Östhammar with a site at Laxemar in Oskarshamn (SKB 2009). Both communities strongly support the placing of a deep-mined geologic repository within their boundaries. SKB thus could focus its decision-making solely on the following technical factors:

- Safety-related site characteristics
- Technology for execution, i.e. the prospects for carrying out the project as robustly and efficiently as possible
- Occupational health and environmental impact
- Societal resources.

SKB did not accord these considerations equal weight. "In the event the analyses do not show a clear difference between the sites, the site that offers the best prospects for longterm safety is selected..." (SKB 2009, 57).

SKB ultimately chose the Östhammar site because it possessed superior characteristics in comparison to the site at Oscarshamn. According to SKB, important differences in the permeability of fractures in the bedrock and somewhat smaller differences in the projected future composition of the groundwater created significant differences in the safety assessments of the two sites. Further, significant future reduction in the salinity of the groundwater at either site could result in degradation of the bentonite, raising the possibility of sulphide-induced corrosion of the copper waste packages. Because groundwater flow is considerably lower at Östhammar, fewer packages would be damaged. At the margins, Östhammar also was preferred because constructing a repository would be easier, and the environmental impacts would be lower. No political ratification of SKB's decision to base its license application on the site located in Östhammar was required.

### *United States*

Under the NWPA, DOE was required to recommend to the President a suite of sites for detailed characterization. After a long and contentious public process, DOE issued site-suitability guidelines (DOE: 1984). It prepared environmental assessments for five potential sites. The analyses contained a common chapter that presented rankings of each site in relation to the guidelines. Initial attempts to aggregate the rankings were sharply criticized by the NAS (Parker 1985a). DOE then decided to use a decision-aiding methodology, multiattribute decision analysis (MUA), in the hope of obtaining greater agreement on how the five sites would be down-selected to three (DOE 1986a [published report]). A second peer-review by the NAS strongly supported the use of the methodology but declined to address the ultimate ranking or recommendation of specific sites for characterization (Parker 1985b). Simultaneously with the release of the MUA report, the Secretary of Energy determined that Yucca Mountain, a salt site in Texas, and a basalt site in Washington would be investigated (DOE 1986b).

The three sites chosen by the Secretary were not the top three identified by the MUA. Considerable public controversy arose, with some parties accusing DOE of manipulating the technique to produce pre-ordained outcomes. In response, DOE stated that the MUA was a "decision-aiding" not a "decision-making" methodology. For example, the methodology did not capture considerations, such as geologic diversity, that the implementer might consider crucial. Of equal importance was the exclusion from the MUA of other considerations, including the degree of local opposition to a repository, risk perceptions associated with transportation of HLW and SNF, geographic equity, and licensability (Merkhofer and Keeny 1987). In short, even the most sophisticated methodology for selecting a site for a deep-mined geologic repository is unlikely to produce agreement if local or state governments believe that the choice is being imposed on them.

## Political Ratification

### *Canada*

Under the Canadian approach of Adaptive Phased Management, volunteer communities are being sought to host a deep-mined geologic repository. If more than one locality accepts and if the technical suitability of the sites is established, a public-engagement program would be launched to consult with interested and affected parties. Safety evaluations and site-specific designs would be prepared. Environmental impact assessments would be developed. Once the Federal Minister of the Environment approves the impact assessments, NWMO would be allowed to apply to the regulators for a Site Preparation License.

### *Finland*

Under the Nuclear Energy Act, an application for a decision-in-principle must be submitted before a site for a deep-mined geologic repository can be approved. The application was filed in 1999 and included statements from the four municipalities under consideration as well as statements from surrounding municipalities. All of the statements were strongly supportive. In addition, the regulator performed preliminary safety assessments. A public hearing then was held. Following the hearing, the Council of State rendered the decision-in-principle in 2000 agreeing to the selection of the site in Eurajoki. The decision-in-principle was confirmed by the Parliament the following year.

### *France*

Although only locations in the Meuse/Haute-Marne region have been studied, a site there for a deep-mined geologic repository cannot be selected simply by default. The 1991 legislation required that public hearings be held on the progress made in each of the four research areas being pursued by ANDRA (OPECST 2005). Subsequently, the Parliament took up the issue and in 2006 passed the Sustainable Management of Radioactive Materials and

Waste Act. Among other things, the law authorizes additional studies directed toward selecting a site in the vicinity of the village of Bure. By 2005, ANDRA had designated a 250 km$^2$ area "transposition zone" within which it believed a deep-mined geologic repository could be constructed. Recently ANDRA identified a 30 km$^2$ area where it will propose to construct the repository's underground works. Discussions are ongoing with the local communities to determine where the repository's surface facilities would be built. Parliament will reach a decision about the site in 2016 after an organized public debate.

### *Japan*

Because no volunteer community has come forward yet, NUMO has not outlined in detail how it intends to select a site for a deep-mined geologic repository if several candidate locations are deemed suitable. The expectation, however, is that an environmental impact assessments would be prepared. Under the Final Disposal of Specific Radioactive Wastes Act, after the Cabinet consents to NUMO's recommendation, the Minister of Economy, Trade, and Industry would need to give formal approval.

### *Switzerland*

Site selection is Stage Three of the *Sectoral Plan*. In that stage, NAGRA would investigate candidate sites in greater detail. With the involvement of the siting region, socioeconomic impacts would be studied. Any benefits package that must be provided would be negotiated and made transparent. Stage Three culminates in the preparation of a general license to construct the deep-mined geologic repository and an environmental impact assessment. The Federal Council would determine whether the general license should be approved; if it is, the Council's decision would have to be ratified by the Parliament. A national referendum may be called, the outcome of which could overturn the parliamentary action.

### *United Kingdom*

Under the Managing Radioactive Waste Safely program, decision-making bodies in local communities would decide whether to proceed to each step in the site-selection process. That determination would be made based on the technical investigations carried out by the British Geological Society as well as the outcome of negotiations over a benefits package. Government would make the final decision about which technically suitable site to choose if more than one locality agrees to host a deep-mined geologic repository, or it would

140      United States Nuclear Waste Technical Review Board

affirm (or reject) the selection of a site if only one community reaches the end of the process.

### *United States*

Although only the site at Yucca Mountain has been characterized, the NWPA still requires that a political process be followed before the site can be officially selected. That process unfolded in 2002. At the beginning of the year, the Secretary of Energy recommended to President George W. Bush that Yucca Mountain be approved for development of a deep-mined geologic repository. The President approved this recommendation the next day and notified Congress. Under the law, the Governor of Nevada had 90 days to veto the selection of the site, which he did. Congress held hearings during the spring and passed a Joint Resolution that overrode the Nevada veto and officially selected Yucca Mountain as the site of a deep-mined geologic repository (U.S. Congress 2002).

## Major Summary Points

- Selecting a site for a deep-mined geologic repository is likely to be less controversial in countries that rely on a voluntarist approach for identifying candidate sites.
- Although the implementer is always responsible for proposing where a deep-mined geologic repository might be located, in most countries, some form of political ratification is required at the national level before a site for a deep-mined geologic repository can be selected.

## APPROVAL TO CONSTRUCT A DEEP-MINED GEOLOGIC REPOSITORY

The processes involved in obtaining approval to construct a deep-mined geologic repository are as varied as the processes involved in identifying candidate sites. In most countries, a representative body, such as the legislature or the Government, makes the final decision. Typically, that body relies on the regulators' advice. In some countries, however, the regulators make the final determination of whether the proposed repository system complies with established requirements. The discussion below of arrangements established in several countries describes the range of variation.

## Canada

Under the current law, the Canadian Nuclear Safety Commission (CNSC) would become involved only when an application for a Site Preparation License is submitted, although the regulators expect to provide comments to NWMO and the public during the siting process. It is entirely up to the implementer to determine the appropriate methodology for demonstrating that dose constraints and risk limits can be met. Scoping assessments, bounding assessments, and realistic calculations are all methodologies that the regulator finds acceptable.

The assessments can be either deterministic or probabilistic (CSNC 2006). Under the Nuclear Safety and Control Act, the CNSC must hold public hearings before making its decision. Beyond that requirement, the regulators' decision-making process is both completely internal and final.

## Finland

STUK has been actively involved in the process that will likely culminate in the submittal of a license application by Posiva in the next few years. It reviewed Posiva's preliminary safety assessments in 1996, 2006, and 2008 as well as the implementer's siting program in 2001.

STUK also has conducted periodic evaluations of the implementer's research and development plans.

STUK requires that Posiva demonstrate the long-term safety of a deep-mined geologic repository by means of a safety analysis. That analysis must address expected evolutions and unlikely disruptive events. Numerical calculations may be complemented by qualitative expert judgment whenever the quantitative analyses are not feasible or are too uncertain. The safety assessment may be either deterministic or probabilistic.

Uncertainties and their importance to safety are assessed in separate analyses (Finnish Council of State 2008). No requirements for public involvement in the regulators' deliberations have been established. STUK only makes recommendations to Government, which holds the final authority to issue a license.

# France

According to the 2006 Radioactive Materials and Waste Planning Act, following the public debate and selection of a site, ANDRA will prepare and submit a license application to construct a deep-mined geologic repository. The application will be reviewed by ASN and the CNE. ASN will be aided in its review by a technical support organization, the Institute for Radiation Protection and Nuclear Safety. The ASN requires ANDRA to demonstrate compliance through the deterministic evaluation of several nominal and disruptive scenarios. In addition, deterministic sensitivity calculations are used to evaluate the effect of uncertainty on repository performance.

The passage of the Transparency and Nuclear Security Act in 2006 gave new powers to communities near nuclear power and fuel cycle plants. Local information committees (CLIS) were given the right to obtain information on the operation and management of those facilities. Armed with that information, the CLIS also will comment on ANDRA's license application, as will the municipalities, districts, and regions affected by the repository project.

The views of ASN, CNE, and the CLIS are sent to OPECST, which, in turn, conveys them to the relevant committees of the National Assembly and the Senate. Subsequently, Government must table a bill prescribing relevant reversibility conditions. Once that law is passed, the license application may be granted by State Council decree after another public debate is held. The application cannot be granted if the satisfaction of the reversibility decree "is not guaranteed."

# Germany

The implementer, the Federal Office of Radiation Protection, would submit an application—including a safety evaluation and an environmental assessment—to the licensing authority of the Land (State) where the proposed repository would be located. The safety evaluation would use deterministic calculations. These calculations would rely on realistic models that use, for example, median values of input parameters. The Nuclear Licensing Procedures Ordinance specifies how the public can be involved. Relevant technical documents must be made available and interested parties have the right to raise objections in writing. The Länder (States) would hold the initial authority for Plan Approval, the procedure that governs whether such a deep-mined geologic repository can be constructed. The federal Ministry for the

Experience Gained from Programs ... 143

Environment, Nature Conservation, and Nuclear Safety, however, has the right to instruct the Plan Approval Authority to issue a license. Decisions by the Länder can be appealed in court by both the applicant and opponents to the facility.

## Sweden

Permits issued under two pieces of legislation, the Nuclear Activities Act and the Environmental Code, are required before a deep-mined geologic repository can be constructed.[44] Both laws instruct SKB to prepare a safety evaluation and an environmental assessment. The evaluation can be developed using a variety of methodologies; deterministic and probabilistic approaches are equally acceptable. SKB submitted its license application in March 2011.

SSM will process the license application under the provisions of the Nuclear Activities Act and submits its views to Government. At roughly the same time, an Environmental Court will review the application under the Environmental Code, holds a public hearing, and issues a statement of comment for both the Government and the host-community. Government will then ask the municipalities that would host the facilities whether they are prepared to accept them. If the municipalities agree, Government decides whether the application is "permissible" under both laws. SSM grants a license under the Nuclear Activities Act and the Radiation Protection Act, which may include stipulations and conditions. The Environment Court grants a license under the Environmental Code, which can also include stipulations and conditions.

Public participation is secured in various ways. Although SSM relies mostly on written statements from interested and affected parties, it can hold hearings if it so chooses. The proceedings of the Environmental Court are open, and interested and affected parties can attend and make statements.

## United Kingdom

The disposal of HLW and SNF in a deep-mined geologic repository is currently regulated by the Environment Agency under the Radioactive Substances Act of 1993.[45] Depending on the enactment of new legislation that is currently under consideration, two potential routes for reaching a regulatory determination on authorizing a deep-mined geologic repository are conceivable. Under a procedure based solely on a "process of agreement," the regulator would provide advice during the period that the implementer is developing the facility. However, regulatory control would start after repository construction but before waste emplacement. Under a staged regulatory process, control would begin as soon as underground investigations start (Environment Agency 2009). At various steps along the repository development process, the regulators would issue permits and licenses. Although guidance has been issued, no decision has been made about what the implementer must do to demonstrate compliance with established dose constraints or risk limits.

Although it is expected that the implementer would engage in continual dialogue with potential host communities, the regulator also would be involved. Proactive efforts would be devised to engage interested and affected parties. Ultimately, the final decision would rest with the regulators, who have broad discretion to accept or reject outside input in order not to "... compromise [their] regulatory independence" (Environment Agency 2009, 36).

## United States

NRC has erected an elaborate structure for determining whether a license to construct a deep-mined geologic repository should be issued. Not only has NRC established highly prescriptive technical requirements, but it also has set forth an intricate process that formally begins once a license application has been submitted. Yet the process in the United States probably offers more opportunities than any other country for interested and affected parties to participate meaningfully in regulatory decision-making.

The regulatory staff first determines whether the application contains sufficient information to permit a meaningful review (NRC 1989). It then begins to evaluate the safety claims advanced by DOE.

In particular, it scrutinizes the technical basis for the safety assessment, which must include calculating the long-term performance of the deep-mined geologic repository using a probabilistic methodology.

In parallel, an independent licensing board conducts an adjudicatory hearing. The hearing format gives admitted parties the right to advance contentions that document disagreements with DOE's claims.[46] Further, the format allows those parties to present witnesses, to carry out discovery of documents held by other parties, and to cross-examine witnesses called by other participants. The licensing board is required to base its decision on whether a license should be granted solely on the information brought forward at the hearing. The board's decision can be appealed to the full Commission and to the courts.

## Major Summary Points

- In most national programs, the regulatory requirements for determining whether the construction of a deep-mined geologic repository should be permitted are typically not prescriptive. The United States is an exception to this general rule.
- Final decisions about permitting the construction of a deep-mined geologic repository can be made by Government or by the regulators, depending on country-specific circumstances.
- National waste-management programs offer varying opportunities for interested and affected parties to influence the regulatory process.

## CONCLUSION

In each of the 13 national programs considered in this report, the long-term management of HLW and SNF has proven more complicated and protracted than initially expected. What was formerly viewed as a relatively simple technical task is now recognized as a complex socio-technical problem involving political negotiations and institutional design challenges as well as path-breaking scientific and technical analyses. Nonetheless, several national programs already have made considerable progress. Sites for a deep-mined geologic repository for HLW and SNF have been selected in four countries— Finland, France, Sweden, and the United States. License applications to construct such a facility have been submitted in two of those nations (the U.S. and Sweden). Applications are likely to be submitted in the other two within the next few years.

The information contained in this report suggests several important conclusions about processes used to develop a deep-mined geologic repository.

- It is possible to elaborate a disposal concept and to advance a safety case, including quantitative performance assessments, that are widely credible not only to scientific and technical communities but also to broad segments of the general population and political leaders. It appears as if a deep-mined geologic repository can be developed in a number of different hydrogeologic environments. An open and transparent technical assessment process, including international peer reviews, increases public and political support.

- It is possible to find communities that are willing to host a deep-mined geologic repository. From the experience gained in countries where sites have been selected, it appears that some communities do so because of their familiarity with other nuclear activities; others do so because of the economic benefits that might accrue in the future. All of those communities, however, were given a meaningful say in the site-selection process. And all of those communities came to be convinced by the respective implementers that the facility could be operated safely.[47]

- Although national programs differ in terms of what is considered an acceptable risk and how to demonstrate whether a deep-mined geologic repository satisfies those standards, international views on these matters are converging. At least for the first few thousands of years after repository closure, dose constraints across countries are within a factor of three of each other and risk limits are within a factor of ten. Only for compliance periods on the order of 100,000 or 1,000,000 years has no international consensus yet been formed on dose constraints, risk limits, and methodology.

- Organizational forms differ significantly across countries, but successful ones have several characteristics in common. Nuclear industry-owned corporations have been successful in Sweden and Finland. A government agency has been successful in France. Rather than organizational form per se, what appears to be important are organizational behaviors, such as leadership continuity, funding stability, and the capacity to inspire public trust and confidence over long periods of time.

Today, more than a half-century after electricity was first produced by splitting the atom, the beneficiaries of that energy source have committed themselves to finding ways to manage the radioactive wastes thereby created in a technically defensible and socially acceptable way. That commitment should be a source for optimism, not only for the generation that produced the wastes, but for succeeding generations as well.

## REFERENCES

Abraham, Spencer. 2002. "Recommendation by the Secretary of Energy Regarding the Suitability of the Yucca Mountain Site for a Repository Under the Nuclear Waste Policy Act of 1982."

Advisory Panel on Alternative Means for Financing and Managing Radioactive Waste Facilities (AM-FM). 1984. *Managing Radioactive Waste—A Better Way.*

Ähagen, Harold et al. 2006. "Approaching a Decision on Final Disposal: The View of the Municipality of Oskarshamn on the Roles of Key Parties and Other Expectations on Their Contributions During the Licensing Phase," in Kjell Andersson, ed., *Proceedings of VALDOR 2006,* pp. 407-413.

Arbeitskreis Auswahlverfahren Endlagersuche (Committee on a Site Selection Procedure for Repository Sites) (AkEnd). 2002. *Site selection procedure for repository sites: Recommendations of the AkEnd.*

ANDRA. 2009. "Financing Radioactive Waste Management." http://www.andra.fr/ international/pages/en/menu21/waste-management/ financing-of-radioactive-wastemanagement-1623.html (Accessed No vember 2010.)

Atomic Energy Commission (AEC). 1968. "Hanford's Highly Radioactive Waste Management Program." AEC 180/30.

Atomic Energy Commission (AEC). 1970. "Policy Relating to the Siting of Fuel Reprocessing Plants and Related Waste Management Facilities." 10 *Code of Federal Regulations,* Part 50, Appendix F.

Atomic Energy of Canada Limited (AECL). 1994. *Environmental Impact Statement on the Concept for Disposal of Canada's Nuclear Waste.* AECL-10711.

Barth, Yannick. 2009. "Framing Nuclear Waste as a Political Issue in France," *Journal of Risk Research,* 12:7, 941-954.

Bundesministerium für Umwelt, Naturschutz, und Reaktorsicherheit (BMU), 2010. "Safety Requirements Governing the Final Disposal of Heat-Generating Radioactive Waste."

Canadian Environmental Assessment Agency (CEAA). 1998. *Nuclear Fuel Waste Management and Disposal Concept: A Report of the Nuclear Fuel Waste Management and Disposal Concept Environmental Assessment Panel.* (Seaborn Report)

Canadian Nuclear Safety Commission (CNSC). 2006. "Assessing the Long Term Safety of Radioactive Waste Management." Regulatory Guide G-320.

148 United States Nuclear Waste Technical Review Board

Carter, Luther. 1987. *Nuclear Imperatives and Public Trust: Dealing with Radioactive Waste.* (Washington DC: Resources for the Future).

Chilvers, J. and J Burgess. 2008. "Power relations: the politics of risk and procedure in nuclear waste governance." *Environment and Planning A 40:* 1881-1900.

Chung, J.B. and H-K Kim 2009. "Competition, economic benefits, trust, and risk perception in siting a potentially hazardous facility." *Landscape and Urban Planning 91:* 8-16.

Committee on Radioactive Management (CoRWM). 2006. *Managing our Radioactive Waste Safely: CoRWM's Recommendations to Government.* CORWM-700.

Cotton, M. 2009. "Ethical assessment in radioactive waste management: a proposed reflective equilibrium-based deliberative approach." *Journal of Risk Research 12:* 603-618.

Cvetkovich, George and Ragnar Löfstedt (eds). 1999. *Social Trust and the Management of Risk.* (London: Earthscan).

Department of Energy (DOE). 1980. *Final Environmental Impact Statement: Management of Commercially Generated Radioactive Waste.* DOE-EIS-0046F. (Washington, DC: Department of Energy).

Department of Energy (DOE). 1984. "General Guidelines for the Preliminary Screening of Potential Sites for a Nuclear Waste Repository." 10 *Code of Federal Regulations,* Part 960.

Department of Energy (DOE). 1986a. *A Multiatrribute Utility analysis of Sites Nominated for Characterization for the First Radioactive-Waste Repository—A Decision-Aiding Methodology.* DOE-RW-0074. (Washington, DC: Office of Civilian Radioactive Waste Management).

Department of Energy (DOE). 1986b. *Recommendation by the Secretary of Energy of Candidate Sites for Site Characterization for the First Radioactive-Waste Repository.* DOE-S-0048. (Washington DC: Office of Civilian Radioactive Waste Management).

Department of Energy (DOE). 2001. *Alternative Means of Financing and Managing the Civilian Radioactive Waste Management Program.* DOE-RW-0546. (Washington, DC: Department of Energy).

Department of Energy (DOE). 2002. *Final Environmental Impact Statement for a Geologic Repository for the Disposal of Spent Nuclear Fuel and High-Level Radioactive Waste at Yucca Mountain, Nye County, Nevada.* DOE/EIS-0250

Environment Agency and Northern Ireland Environment Agency. 2009. *Geologic Disposal Facilities on Land for Solid Radioactive Wastes: Guidance on Requirements for Authorisation.*

Environmental Protection Agency (EPA). 2005. Letter to Charles J. Fitzpatrick. "Response to Freedom of Information Act Appeal HQ-RIN-00375-05-A." (May 11, 2005).

Environmental Protection Agency (EPA). 2006. "Criteria for the Certification and Recertification of the Waste Isolation Pilot Plant's Compliance with the Disposal Regulations: Recertification Decision," *Federal Register,* 71, (April 10, 2006), pp. 18010-18021.

Federal Office of Radiation Protection (BfS). 2010. http://www.bfs.de/en/endlager/gorleben. (Accessed November 2010.)

Finnish Council of State. 2008. "Government Decision on the Safety of Disposal of Spent Nuclear Fuel."

Finnish Energy Industries. 2007. "Nuclear Waste Management in Finland." Harrington, John. 1985. Letter to President George Bush dated April 18, 1985.

Hedin, A. 2008. "Safety Functions and Safety Function Indicators—Key Elements in SKB's Methodology for Assessing Long-Term Safety of a KBS-3 Repository," in in Nuclear Energy Agency, *Safety Cases for Deep Geological Disposal of Radioactive Waste: Where Do We Stand?* NEA-6319. (Paris: Organization for Economic Cooperation and Development), pp. 221-230.

Hocke, Peter and Ortwin Renn. 2010. "Concerned Public and the Paralysis of Decision-Making: Nuclear Waste Management Policy in Germany," *Journal of Risk Research,* 12, 7, 921-940.

Herzik, Eric B., and Alvin H. Mushkatel. 1993. *Problems and Prospects for Nuclear Waste Disposal Policy.* Westport, CT: Greenwood Press.

Interagency Review Group on Nuclear Waste Management (IRG). 1978. *Subgroup Report on Alternative Technology Strategies of the Isolation of Nuclear Waste.* TID-28818 (Draft). (Washington DC: Department of Energy).

Interagency Review Group on Nuclear Waste Management (IRG). 1979. *Report to the President by the Interagency Review Group on Nuclear Waste Management.* TID29442. (Washington DC: Department of Energy).

International Atomic Energy Agency (IAEA). 2007. *Factors Affecting Public and Political Acceptance for the Implementation of Geological Disposal.* IAEA-TECDOC-1566. (Vienna: International Atomic Energy Agency).

## 150     United States Nuclear Waste Technical Review Board

Jacob, Gerald. 1990. *Site Unseen: The Politics of Siting a Nuclear Waste Repository.* (Pittsburgh: University of Pittsburgh Press.)

Jonas, Hans. 1973. "Technology and Responsibility: Reflections on the New Task of Ethics," *Social Research,* 40, Spring, 1973.

Kärnavfallsfonden. 2010. *Annual Report 2009.*

Kasperson, Roger, Dominic Golding, and Seth Tuler. 1992. "Social Distrust as a Factor in Siting Hazardous Facilities and Communicating Risk," *Journal of Social Issues,* 48.

Krone, J. et al. 2008. "Developing an Advance Safety Concept for an HLW Repository in Salt Rock," in Nuclear Energy Agency, *Safety Cases for Deep Geological Disposal of Radioactive Waste: Where Do We Stand?* NEA-6319. (Paris: Organization for Economic Cooperation and Development), pp. 195-202.

Krütlia, P et al. 2010. "Functional-dynamic public participation in technological decision-making: site selection processes of nuclear waste repositories." *Journal of Risk Research,* DOI: 10.1080/1366987 1003703252.

Lee, Kai. 1999. "Appraising Adaptive Management," *Conservation Ecology,* 2.

Lindblom, Charles. 1959. "The Science of Muddling Through," *Public Administration Review,* (Spring), pp. 79-88.

Macdonald, D. and Samin Sharifi-Asi. 2009. *Is Copper Immune to Corrosion When in Contact with Water and Aqueous Solutions?* Swedish Radiation Safety Authority, Research Report 2011:09.

Merkhofer, Miley and Ralph Keeny. 1987. "A Multiattribute Utility Analysis of Alternative Sites for the Disposal of Nuclear Waste," *Risk Analysis,* 7, June 1987, pp. 173-194.

Miller, Bob. 1996. Letter to Hazel O'Leary, Secretary of Energy. December 24. 1996.

Müller-Hoeppe et al. 2008. "The Role of Structural Reliability of Geotechnical Barriers of an HLW/SNF Repository in Salt Rock Within the Safety Case," in Nuclear Energy Agency, *Safety Cases for Deep Geological Disposal of Radioactive Waste: Where Do We Stand?* NEA-6319. (Paris: Organization for Economic Cooperation and Development), pp. 231-239.

National Academy of Sciences (NAS). 1957. *The Disposal of Radioactive Waste on Land.* (Washington DC: National Academy of Sciences).

National Academy of Sciences (NAS). 1983. *A Study of the Isolation System for Geologic Disposal of Radioactive Wastes.* (Washington DC: National Academy of Sciences).

National Academy of Sciences (NAS). 1995. *Technical Basis for Yucca Mountain Standards.* (Washington DC: National Academy of Sciences)

National Academy of Sciences (NAS). 2003. *One Step at a Time: The Staged Development of Geologic Repositories for High-Level Waste.* (Washington DC: National Academy of Sciences).

National Cooperative for the Disposal of Radioactive Waste (NAGRA). 2002. *Project Opalinus Clay: Safety Report.* Technical Report 02-05.

Nuclear Energy Agency (NEA). 2001. *An International Peer Review of Safety Report 97: Post-closure Safety of a Deep Repository for Nuclear Spent Fuel in Sweden.* (Paris: Organization for Economic Cooperation and Development.

Nuclear Energy Agency (NEA). 2004a. *Stepwise Approach to Decision Making for Longterm Radioactive Waste Management—Experience, Issues, and Guiding Principles.* NEA-4429. (Paris: Organization for Economic Cooperation and Development).

Nuclear Energy Agency (NEA). 2004b. *Learning and Adapting to Societal Requirements for Radioactive Waste Management.* NEA-5296. (Paris: Organization for Economic Cooperation and Development).

Nuclear Energy Agency (NEA). 2004c. *Safety of Disposal of Spent Fuel, HLW, and Long-lived ILW in Switzerland.* NEA-5568. (Paris: Organization for Economic Cooperation and Development).

Nuclear Energy Agency (NEA). 2008a. *Safety Cases for Deep Geological Disposal of Radioactive Waste: Where Do We Stand?* NEA-6319. (Paris: Organization for Economic Cooperation and Development).

Nuclear Energy Agency (NEA). 2008b. "Moving Forward with Geological Disposal of Radioactive Waste: An NEA RWMC Collective Statement." NEA-6433. (Paris: Organization for Economic Cooperation and Development).

Nuclear Energy Agency (NEA). 2009. *Towards Transparent, Proportionate, and Deliverable Regulation for Geologic Disposal.* NEA/RWM/RF2009(1). (Paris: Organization for Economic Cooperation and Development).

Nuclear Energy Agency (NEA). 2010. *Partnering for Long-Term Management of Radioactive Waste.* NEA-6823. (Paris: Organization for Economic Cooperation and Development).

Nuclear Regulatory Commission (NRC). 1989. "Procedures Applicable to Proceedings for the Issuance of Licenses for the Receipt of High-Level Radioactive Waste at a Geologic Repository." 10 *Code of Federal Regulations* Part 2, Subpart J.

152     United States Nuclear Waste Technical Review Board

Nuclear Regulatory Commission. 2001. "Part 63—Disposal of High-Level Radioactive Wastes in a Geologic Repository at Yucca Mountain, Nevada." *Federal Register*, 66, 55792.

Nuclear Waste Management Organization of Japan (NUMO). 2004. *Evaluating Site Suitability for a HLW Repository, Scientific Background and Practical Application of NUMO's Siting Factors.* NUMO-TR-04-04. (Tokyo: Nuclear Waste Management Organization of Japan).

Nuclear Waste Management Organization (NWMO). 2005. *Choosing a Way Forward: The Future Management of Canada's Used Nuclear Fuel,* Final Study.

Nuclear Waste Management Organization (NWMO). 2007. *Moving Forward Together— Annual Report 2007.*

Nuclear Waste Management Organization (NWMO). 2009. *Moving Forward Together: Designing the Process for Selecting a Site.*

Nuclear Waste Technical Review Board (NWTRB). 2008. *Report to the U.S. Congress and the Secretary of Energy: March 1, 2006—December 31, 2008.* (Arlington VA: U.S. Nuclear Waste Technical Review Board).

Nuclear Waste Technical Review Board (NWTRB). 2009. *Survey of National Programs for Managing High-Level Radioactive Waste and Spent Nuclear Fuel.* (Arlington VA: U.S. Nuclear Waste Technical Review Board).

Office for Scientific and Technological Options (OPECST). 2005. *Looking after the Longer Term: An Act in 2006 on the Sustainable Management of Radioactive Wastes.*

Office of Technology Assessment (OTA). 1985. *Managing the Nation's Commercial High-Level Radioactive Waste.* OTA-O-171. (Washington, DC: U.S. Government Printing Office).

Parker, Frank. 1985a. Letter to Ben Rusche, Director, Office of Civilian Radioactive Waste Management. (April 26, 1985).

Parker, Frank. 1985b. Letter to Ben Rusche, Director, Office of Civilian Radioactive Waste Management. (October 10, 1985).

Posiva Oy. 1999. *The Final Disposal Facility for Spent Nuclear Fuel: Environmental Impact Assessment Report.* (Helsinki: Posiva Oy).

Secretary of Energy Advisory Board (SEAB). 1993. *Earning Public Trust and Confidence: Requisites for Managing Radioactive Wastes.* (Washington DC: U.S. Department of Energy).

Sjoberg, L. 2009. "Precautionary attitudes and the acceptance of a local nuclear waste repository." *Safety Science.* 47: 542-546.

Swedish National Council for Nuclear Waste. 2007. *Final Disposal of Spent Nuclear Fuel— Regulatory System and Roles of Different Actors during the Decision Process.*

Swedish National Council for Nuclear Waste. 2008. *Final Disposal of Nuclear Waste: The Swedish National Council for Nuclear Waste's Review of the Swedish Nuclear Fuel and Waste Management Co's. (SKB's) RD&D Programme 2007.* SOU 2008:70.

Swedish National Council for Nuclear Waste. 2009. *Mechanisms of Copper Corrosion in Aqueous Environments.* Report 2009:4e.

Swedish Nuclear Fuel Supply Company (SKB). 1978a. *Handling of Spent Nuclear Fuel and Final Storage of Vitrified High Level Reprocessing Waste.* (Stockholm: Swedish Nuclear Fuel Supply Company).

Swedish Nuclear Fuel Supply Company/Division KBS (SKB). 1978b. *Handling and Final Storage of Spent Nuclear Fuel—KBS-3.* (Stockholm: Swedish Nuclear Fuel Supply Company).

Swedish Nuclear Fuel Supply Company/Division KBS (SKB). 1983. *Final Storage of Spent Nuclear Fuel—KBS-3.* (Stockholm: Swedish Nuclear Fuel Supply Company).

Swedish Nuclear Fuel and Waste Management Company (SKB). 2009. *Final Repository for Spent Nuclear Fuel in Forsmark—Basis for Decision and Reasons for Site Selection.*

Swiss Federal Office of Energy (SFOE). 2008. *Sectoral Plan for Deep Geological Repositories: Conceptual Part.*

Swiss Federal Office of Energy (SFOE). 2009. "Decommissioning and Waste Disposal Funds." http://www.bfe.admin.ch/entsorgungsfonds/ index.html? lang=en. (Accessed November 2010.)

Sundqvist, Göran. 2002. *The Bedrock of Opinion: Science, Technology, and Society in the Siting of High-Level Nuclear Waste.* (Dordrecht: Kluwer Academic Publishers).

Thomas, Craig. 1993. "AM-FM's Corporate Solution for Radioactive Waste Management: Appealing But Appropriate?" in *Compilation of Reports Prepared for the Secretary of Energy Advisory Board Task Force on Radioactive Waste Management.* (Washington DC: Department of Energy), pp. 261-305.

Thompson, James and Arthur Tuden. 1959. "Strategies, Structures, and Processes of Organizational Decision-Making," in James Thompson et al., *Comparative Studies in Administration,* (Pittsburgh: University of Pittsburgh Press). 195-216.

## 154 United States Nuclear Waste Technical Review Board

United Kingdom Government (UKG). 2006. *Response to the Report and Recommendations from the Committee on Radioactive Waste Management (CoRWM) By the UK Government and the Devolved Administrations.* (Norwich: Department of Environment, Food, and Rural Affairs).

United Kingdom Government (UKG). 2008. *Managing Radioactive Waste Safely: A Framework for Implementing Geological Disposal.*

U.S. Congress. 1963. Joint Committee on Atomic Energy, hearings, *Chemical Reprocessing Plant,* 88th Congress, 1st Session, (Washington, DC: U.S. Government Printing Office).

U.S. Congress. 1987. "Establishing a Nuclear Waste Policy Review Commission, an Office of the Nuclear Waste Negotiator, and for Other Purposes," *Report from the U.S. House of Representatives, Committee on Interior and Insular Affairs.* Report 100-425.

U.S. Congress. 2002. "Joint Resolution Approving the Site at Yucca Mountain, Nevada, for the Development of a Repository for the Disposal of High-Level Radioactive Waste and Spent Nuclear Fuel, Pursuant to the Nuclear Waste Policy Act of 1982." Public Law 107-200.

Vaughn, Diane. 1996. *The Challenger Launch Decision: Risky Technology, Culture, and Deviance at NASA.* (Chicago: University of Chicago Press).

Westinghouse Waste Isolation Division. 1995. *Waste Isolation Pilot Plant Safety Analysis Report.* DOE/WIPP-95-0265.

World Nuclear Association (WANO). 2010. "Nuclear Power in South Korea," http://www. world-nuclear.org/info/inf81.html. (Visited in November 2010.)

## End Notes

[1] In some countries, HLW and SNF also are produced as result of nuclear-weapons and nuclear-propulsion programs and as a result of experimental and research reactor programs. Typically, the requirements for the long-term disposition of those wastes are no different from the requirements for disposing of commercially generated waste. No further distinction will be made in this report among the various sources of high-activity radioactive waste.

[2] The discussion in this section focuses on the early days of radioactive waste management policy-making in the United States. A similar tale, however, could be told of the development of policies in the other countries that had generated radioactive waste by the end of the 1960s including the United Kingdom, France, and the Union of Soviet Socialist Republics.

Experience Gained from Programs ...   155

[3] The countries are the United States, Belgium, Canada, China, Finland, France, Germany, Japan, the Republic of Korea, Spain, Sweden, Switzerland, and the United Kingdom. This report focuses entirely on the efforts of those countries to develop programs for the long-term management of HLW and SNF. It does not consider directly associated activities such as long-term storage, transportation of radioactive waste, and initiatives to reprocess SNF or recycle the reprocessed products.

[4] The reader is advised to refer to this Board report for more-detailed information about each national program.

[5] The discussion that follows includes illustrations drawn from various national programs. The illustrations were chosen to represent the range of approaches taken by the 13 countries; they are not meant to be exhaustive.

[6] At least one international organization, the Nuclear Energy Agency's (NEA) Forum on Stakeholder Confidence, was established solely to provide a venue to discuss these issues.

[7] Because all but one of the 13 countries are pluralistic democracies, these questions could not be easily dismissed or branded as illegitimate.

[8] Stretching involves the municipality developing and posing sharply focused questions to the body responsible for developing a deep-mined geologic repository and ensuring that the answers are responsive and specific.

[9] At the moment, the partnership in Belgium is addressing the management of low- and intermediate-level waste. The West Cumbria Partnership in the United Kingdom involves local governments interacting with the national government. The implementer is just observing the process.

[10] For a discussion of why it may be hard to implement recommendations to foster public trust and confidence, see SEAB: 1993, 61.

[11] It is unclear what Germany's position is. That country's Safety Requirements Governing the Disposal of Heat-Generating Radioactive Waste calls for "stepwise optimization." Yet only one license is likely to be issued by the regulators, covering authorization to construct a deep-mined geologic repository, to receive and possess waste, and to decommission the facility.

[12] Neither Belgium nor Spain has officially adopted this option. The current Government in Scotland also is opposed to it, although previous Governments have supported the option of geologic disposal.

[13] The passage of environmental impact legislation has affected how waste-management options are chosen. Although the option of disposal in a deep-mined geologic repository was, for all practical purposes, selected in the United States in 1970, it was not until 1980 that the environmental assessments necessary to support that choice were completed (DOE 1980).

[14] In Germany, perhaps more than any other country, the fate of national program for managing HLW and SNF is closely tied to the ongoing debate over the future of nuclear power.

[15] An amendment to the Act was passed in early 1998 to allow funding for a fourth major research area, the evaluation of interim storage options.

[16] The Detailed Tables in NWTRB 2009 provide additional information about the organizations that are part of each national waste-management program.

[17] The Division of Industrial Participation was a unit within the AEC, charged with expanding commercial opportunities for the private sector. Its limited influence made it no match for the Reactor Development Division, which championed the option of having a government agency be responsible for developing a deep-mined geologic repository.

[18] In Germany, the 4th Amendment to the Atomic Energy Act of 1976, which assigned the task of waste disposal to the Federal Government, stipulated that the Government could make use

# 156      United States Nuclear Waste Technical Review Board

of third parties to discharge those duties. Toward this end, the Government created the German Service Company for the Construction and Operation of Waste Repositories (DBE) and entered into an exclusive contract with it. DBE is currently 25 percent state-owned with the balance owned by German nuclear utilities, although state ownership will double shortly. In Korea, the Radioactive Waste Management Company was created by Parliament in 2008. It has been described as an "umbrella organization set up to resolve South Korea's waste management issues and waste disposition, and particularly to forge a national consensus on high-level wastes" (WANO 2010). No further information is available about this implementer's organizational form. Until it has been established and begins operations, responsibilities for waste management are vested in a government agency.

[19] When SKB was formed in 1973, the Swedish State, through its ownership of utilities, was the major nuclear-electricity generator at the time. SKB initially focused on the supply of fuel to the nuclear power plants. By the mid-1970s, the company's focus had shifted to nuclear waste management.

[20] Some of the Canadian, Finnish, and Swedish utilities that own the implementers are partly government-owned.

[21] The AM-FM study, which is the most systematic one to date, still advances impressionistic claims that are only tenuously based on evidence. See Thomas 1993.

[22] In France and the United States, the total cost is divided between the generators/consumers of nuclear electricity and the government, which has to pay to manage waste from defense nuclear programs. In the United Kingdom, the government will pay the costs of disposing of waste from defense nuclear programs and waste from the older operating and decommissioned nuclear power plants. The costs of managing waste from new-build nuclear power plants must be paid by their owners.

[23] In Germany and Belgium, the waste producers reimburse the government.

[24] One mil equals $0.001.

[25] A more technically defensible approach would be to focus on the total hydrogeologic environment rather than rock type.

[26] The viability of any disposal concept ultimately will depend on finding an appropriate site and characterizing the site in sufficient detail.

[27] An excellent resource for understanding the development of disposal concepts is NEA 2008a.

[28] A very similar concept was used by DOE in developing the disposal concept for defense-origin transuranic waste at the Waste Isolation Pilot Plant (WIPP) in New Mexico.

[29] This position is consistent with the safety analysis carried out at WIPP in that the hypothetical releases are dominated by the human intrusion scenario. "Natural" releases are calculated to be minuscule (Westinghouse 1995).

[30] See Table 7 and the Detailed Tables in NWTRB 2009 for a more complete description of the regulations that have been adopted to date.

[31] In the United States, regulators sometimes will publish an "Advanced Notice of Proposed Rulemaking" to solicit the views of interested and affected parties before the release of a draft regulation.

[32] The approach was recommended by NAS (NAS 1995).

[33] The provision involves limits on groundwater travel time from the repository to the accessible environment.

[34] As the IRG recognized, and as will be noted below, there is no unambiguous boundary that differentiates a "technical" from a "nontechnical" filter. At the margins, some overlap and interdependence may, even must, exist.

[35] Importantly, to date, no country uses a whole-system performance assessment to select the initial suite of candidate sites. At subsequent stages in the site-selection process, however, more holistic and systemic approaches are brought to bear.

[36] The recommendations advanced by AkEnd group were never adopted by the German Federal Government. Subsequently, the German Geological Survey concluded that crystalline rock formations in Germany were not suitable for development as a deep-mined geologic repository.

[37] Before the process was changed in 2003, the referendum was held only in the canton where a repository might be sited.

[38] These communities always make clear that their continued interest depends on being convinced that a repository would operate safely.

[39] Only defense-origin transuranic waste can be disposed of at WIPP under present regulations and agreements. However, because that facility is the only operating deep-mined geologic repository, its story may be instructive for efforts to site a facility for HLW and SNF.

[40] Many analysts and commentators, for instance, point to the success of the siting process in Sweden and argue that it should be emulated elsewhere. Although the process may work well in Sweden, the Swedish model has not (yet) been successfully replicated in any of the other countries that have consciously adopted it (IAEA 2007).

[41] The Japanese and the American examples illustrate the so-called "donut effect," in which a local community willing to host a deep geologic repository can be blocked by officials representing governments at the state or regional level.

[42] See, for example, comments on the proposed 10 CFR 960 from the State of Nevada suggesting that oxidizing conditions should disqualify a proposed site and comments from the State of Mississippi seeking to limit the ability of DOE to site a repository in salt domes.

[43] In Sweden, the two detailed investigations had been preceded by feasibility studies in eight communities, resulting in the identification a total of eight sites in five communities as having the potential for detailed investigation.

[44] See, National Council for Nuclear Waste 2007 for a good discussion of the licensing process.

[45] The Nuclear Installations Inspectorate will regulate operational activities at a repository.

[46] It was at this point in the process that NRC's consideration of the Yucca Mountain license application was suspended. It is unclear whether or when the proceedings will resume. Therefore, the discussion that follows reflects only the process that has been put in place.

[47] In Federal systems, such as those found in Japan and the United States, it may be necessary to secure the approval of a politically superior state or prefecture. This requirement may complicate any voluntary process.

In: Spent Nuclear Fuel
Editors: Phillip T. Crawford et al.

ISBN: 978-1-62257-347-9
© 2012 Nova Science Publishers, Inc.

*Chapter 3*

# SPENT FUEL STORAGE IN POOLS AND DRY CASKS: KEY POINTS AND QUESTIONS AND ANSWERS[*]

## *United States Nuclear Regulatory Commission*

### KEY POINTS

1. All US nuclear power plants store spent nuclear fuel in "spent fuel pools." These pools are robust constructions made of reinforced concrete several feet thick, with steel liners. The water is typically about 40 feet deep, and serves both to shield the radiation and cool the rods.
2. As the pools near capacity, utilities move some of the older spent fuel into "dry cask" storage. Fuel is typically cooled at least 5 years in the pool before transfer to cask. NRC has authorized transfer as early as 3 years; the industry norm is about 10 years.
3. The NRC believes spent fuel pools and dry casks both provide adequate protection of the public health and safety and the

---

[*] This is an edited, reformatted and augmented version of the United States Nuclear Regulatory Commission publication, adapted from the following website: http://www.nrc.gov/waste/spent-fuel-storage/faqs.html, dated March 2012

# 160 The United States Nuclear Regulatory Commission

environment. Therefore there is no pressing safety or security reason to mandate earlier transfer of fuel from pool to cask.

4. After the September 11, 2001, terrorist attacks, the NRC issued orders to plant operators requiring several measures aimed at mitigating the effects of a large fire, explosion, or accident that damages a spent fuel pool. These were meant to deal with the aftermath of a terrorist attack or plane crash; however, they would also be effective in responding to natural phenomena such as tornadoes, earthquakes or tsunami. These mitigating measures include:
   1. Controlling the configuration of fuel assemblies in the pool to enhance the ability to keep the fuel cool and recover from damage to the pool.
   2. Establishing emergency spent fuel cooling capability.
   3. Staging emergency response equipment nearby so it can be deployed quickly

5. According to the Congressional Research Service (using NEI data), there were 62,683 metric tons of commercial spent fuel accumulated in the United States as of the end of 2009.
   1. Of that total, 48,818 metric tons – or about 78 percent – were in pools.
   2. 13,856 metric tons – or about 22 percent – were stored in dry casks.
   3. The total increases by 2,000 to 2,400 tons annually.

## QUESTIONS AND ANSWERS OVERVIEW:

- Questions and Answers – General
  - o What is spent nuclear fuel?
  - o Why does spent fuel need to be cooled?
- Questions and Answers – Spent Fuel Inventories
  - o Why doesn't the NRC have up-to-date figures on how much spent fuel is stored at US nuclear plants?
  - o How much fuel is currently in dry cask storage?
  - o How much fuel is stored at decommissioned reactors? Is it in pools or casks?
- Questions and Answers – Spent Fuel Pool Safety
  - o What do you look at when you license a fuel storage facility? How do I know it can withstand a natural disaster?

# Spent Fuel Storage in Pools and Dry Casks    161

- o How do you know the fuel pools are safe? Does the NRC inspect these facilities, or just the reactor itself?
- o What would happen to a spent fuel pool during an earthquake? How can I be sure the pool wouldn't be damaged?
- o Can spent fuel pools leak?
- o How would you know about a leak in such a large pool of water?
- o How can operators get water back in the pool if there is a leak or a failure?
- o Do US nuclear power plants store their fuel above grade? Why is this considered safe?
- o How are spent fuel pools kept cool? What happens if the cooling system fails?
- o What keeps spent fuel from re-starting a nuclear chain reaction in the pool?
- Questions and Answers – ISFSIs
  - o What is dry cask storage?
  - o What is an "ISFSI"?
  - o What kind of license is required for an ISFSI?
- Questions and Answers – Dry Cask Safety
  - o How do you know the dry casks are safe? Does the NRC inspect these facilities, or just the reactor and spent fuel pool?
  - o What keeps fuel cool in dry casks?
- Questions and Answers – Waste Confidence & Future Plans
  - o How long is spent fuel allowed to be stored in a pool or cask?
  - o The most recent waste confidence findings say that fuel can be stored safely for 60 years beyond the reactor's licensed life. Does this mean fuel will be unsafe starting in 2059 [60 years after Dresden 1's original license ended]? What if the spent fuel pool runs out of room even before the end of a reactor license? What is the NRC going to do about this?
  - o Does the waste confidence decision mean that a particular cask is safe?
  - o The waste-confidence revision seems like a long-term effort. What is the NRC doing to improve safety of spent fuel storage now?
  - o The NRC is reviewing applications for new nuclear power plants. What is the environmental impact of all that extra fuel?
- Questions and Answers – Security

## 162 The United States Nuclear Regulatory Commission

- o What about security? How do you know terrorists won't use all of this waste against us?
- Questions and Answers – Emergency Planning
- o What emergency plans are required for spent fuel storage facilities at nuclear power plants undergoing decommissioning or sites that have completed decommissioning?

## QUESTIONS AND ANSWERS – GENERAL

### What Is Spent Nuclear Fuel?

"Spent nuclear fuel" refers to fuel elements that have been used at commercial nuclear reactors, but that are no longer capable of economically sustaining a nuclear reaction. Periodically, about one-third of the nuclear fuel in an operating reactor needs to be unloaded and replaced with fresh fuel.

### Why Does Spent Fuel Need to Be Cooled?

Spent fuel continues to generate heat because of radioactive decay of the elements inside the fuel. After the fission reaction is stopped and the reactor is shut down, the products left over from the fuel's time in the reactor are still radioactive and emit heat as they decay into more stable elements. Although the heat production drops rapidly at first, heat is still generated many years after shutdown. Therefore, the NRC sets requirements on the handling and storage of this fuel to ensure protection of the public and the environment.

## QUESTIONS AND ANSWERS – SPENT FUEL INVENTORIES

### Why Doesn't the NRC Have up-to-Date Figures on How Much Spent Fuel Is Stored at US Nuclear Plants?

The NRC and Department of Energy (NNSA) operate the Nuclear Material Management and Safeguards System (NMMSS), a database that tracks Special Nuclear Material (enriched uranium and plutonium). This database does not distinguish between fresh and irradiated material, and the

# Spent Fuel Storage in Pools and Dry Casks

information is withheld from the public for security reasons. That's why figures on spent fuel inventory come from the industry.

## How Much Fuel Is Currently in Dry Cask Storage?

As of November 2010, there were 63 "independent spent fuel storage installations" (or ISFSIs) licensed to operate at 57 sites in 33 states. These locations are shown on a map on the NRC website at: *http://www.nrc.gov/ waste/spent-fuel-storage/locations.pdf*. Over 1400 casks are stored in these independent facilities.

## How Much Fuel Is Stored at Decommissioned Reactors? Is It in Pools or Casks?

There are currently 10 decommissioned nuclear power reactors at 9 sites with no other nuclear operations. According to a 2008 Department of Energy report to Congress, approximately 2800 metric tons of spent fuel is stored at these nine sites. As of the writing of that report, seven of the sites had independent spent fuel storage installations, or ISFSIs. Two additional sites had approximately 1000 metric tons of spent fuel remaining in pool storage.

## QUESTIONS AND ANSWERS – SPENT FUEL POOL SAFETY

### What Do You Look at When You License a Fuel Storage Facility? How Do I Know It Can Withstand a Natural Disaster?

The NRC's requirements for both wet and dry storage can be found in Title 10 of the Code of Federal Regulations (10 CFR), including the general design criteria in Appendix A to Part 50 and the spent-fuel storage requirements in Part 72. The staff uses these rules to determine that the fuel will remain safe under anticipated operating and accident conditions. There are requirements on topics such as radiation shielding, heat removal, and criticality. In addition, the staff reviews fuel storage designs for protection against:

# 164      The United States Nuclear Regulatory Commission

- natural phenomena, such as seismic events, tornados, and flooding
- dynamic effects, such as flying debris or drops from fuel handling equipment and drops of fuel storage and
- handling equipment hazards to the storage site from nearby activities

## How Do You Know the Fuel Pools Are Safe? Does the NRC Inspect These Facilities, or Just the Reactor Itself?

NRC inspectors are responsible for verifying that spent fuel pools and related operations are consistent with a plant's license. For example, our staff inspects spent fuel pool operations during each refueling outage. We also performed specialized inspections to verify that new spent fuel cooling capabilities and operating practices were being implemented properly.

## What Would Happen to a Spent Fuel Pool During an Earthquake? How Can I Be Sure the Pool Wouldn't Be Damaged?

All spent fuel pools are designed to seismic standards consistent with other important safety-related structures on the site. The pool and its supporting systems are located within structures that protect against natural phenomena and flying debris.

The pools' thick walls and floors provide structural integrity and further protection of the fuel from natural phenomena and debris. In addition, the deep water above the stored fuel (typically more than 20 feet above the top of the spent fuel rods) would absorb the energy of debris that could fall into the pool. Finally, the racks that support the fuel are designed to keep the fuel in its designed configuration after a seismic event.

## Can Spent Fuel Pools Leak?

Spent fuel pools lined with stainless steel are designed to protect against a substantial loss of the water that cools the fuel. Pipes typically enter the pool above the level of the stored fuel, so that the fuel would stay covered even if there were a problem with one of the pipes.

The only exceptions are small leakage-detection lines and, at two pressurized water reactor (PWR) sites, robust fuel transfer tubes that enter the spent fuel pool directly.

The liner normally prevents water from being lost through the leak detection lines, and isolation valves or plugs are available if the liner experiences a large leak or tear.

## How Would You Know About a Leak in Such a Large Pool of Water?

The spent fuel pools associated with all but one operating reactor have liner leakage collection to allow detection of very small leaks. In addition, the spent fuel pool and fuel storage area have diverse instruments to alert operators to possible large losses of water, which could be indicated by low water level, high water temperature, or high radiation levels.

## How Can Operators Get Water Back in the Pool if There Is a Leak or a Failure?

All plants have systems available to replace water that could evaporate or leak from a spent fuel pool. Most plants have at least one system designed to be available following a design basis earthquake. In addition, the industry's experience indicates that systems not specifically designed to meet seismic criteria are likely to survive a design basis earthquake and be available to replenish water to the spent fuel pools. Furthermore, plant operators can use emergency and accident procedures that identify temporary systems to provide water to the spent fuel pool if normal systems are unavailable. In some cases, operators would need to connect hoses or install short pipes between systems. The fuel is unlikely to become uncovered rapidly because of the large water volume in the pool, the robust design of the pool structure, and the limited paths for loss of water from the pool.

## Do US Nuclear Power Plants Store Their Fuel Above Grade? Why Is This Considered Safe?

For boiling water reactor (BWR) Mark I and II designs, the spent fuel pool structures are located in the reactor building at an elevation several stories above the ground (about 50 to 60 feet above ground for the Mark I reactors). The spent fuel pools at other operating reactors in the US are typically located with the bottom of the pool at or below plant grade level. Regardless of the location of the pool, its robust construction provides the potential for the structure to withstand events well beyond those considered in the original design. In addition, there are multiple means of restoring water to the spent fuel pools in the unlikely event that any is lost.

## How Are Spent Fuel Pools Kept Cool? What Happens If the Cooling System Fails?

The spent fuel pool is cooled by an attached cooling system. The system keeps fuel temperatures low enough that, even if cooling were lost, operators would have substantial time to recover cooling before boiling could occur in the spent fuel pool. Licensees also have backup ways to cool the spent fuel pool, using temporary equipment that would be available even after fires, explosions, or other unlikely events that could damage large portions of the facility and prevent operation of normal cooling systems. Operators have been trained to use this backup equipment, and it has been evaluated to provide adequate cooling even if the pool structure loses its water-tight integrity.

## What Keeps Spent Fuel from Re-Starting a Nuclear Chain Reaction in the Pool?

Spent fuel pools are designed with appropriate space between fuel assemblies and neutron-absorbing plates attached to the storage rack between each fuel assembly. Under normal conditions, these design features mean that there is substantial margin to prevent criticality (i.e., a condition where nuclear fission would become self-sustaining). Calculations demonstrate that some margin to criticality is maintained for a variety of abnormal conditions, including fuel handling accidents involving a dropped fuel assembly.

# Questions and Answers – ISFSIs

## What Is Dry Cask Storage?

Dry cask storage allows spent fuel that has already been cooled in the spent fuel pool for several years to be surrounded by inert gas inside a container called a cask. The casks are typically steel cylinders that are either welded or bolted closed.

The steel cylinder provides containment of the spent fuel. Each cylinder is surrounded by additional steel, concrete, or other material to provide radiation shielding to workers and members of the public.

## What is an "ISFSI"?

An independent spent fuel storage installation, or ISFSI, is a facility that is designed and constructed for the interim storage of spent nuclear fuel. These facilities are licensed separately from a nuclear power plant and are considered independent even though they may be located on the site of another NRC-licensed facility.

## What Kind of License Is Required for an ISFSI?

NRC authorizes storage of spent nuclear fuel at an ISFSI in two ways: site-specific or general license. For site-specific applications, the NRC reviews the safety, environmental, physical security and financial aspects of the licensee and proposed ISFSI and, if we conclude it can operate safely, we issue a license.

This license contains requirements on topics such as leak testing and monitoring and specifies the quantity and type of material the licensee is authorized to store at the site.

A general license authorizes storage of spent fuel in casks previously approved by the NRC at a site already licensed to possess fuel to operate a nuclear power plant. Licensees must show the NRC that it is safe to store spent fuel in dry casks at their site, including analysis of earthquake intensity and tornado missiles.

Licensees also review their programs (such as security or emergency planning) and make any changes needed to incorporate an ISFSI at their site.

168 The United States Nuclear Regulatory Commission

Of the currently licensed ISFSIs, 48 are operating under general licenses and 15 have specific licenses.

## QUESTIONS AND ANSWERS – DRY CASK SAFETY

### How Do You Know the Dry Casks Are Safe? Does the NRC Inspect These Facilities, or Just the Reactor and Spent Fuel Pool?

The NRC is responsible for inspection of dry cask storage. All casks also undergo a safety review before they are certified for use by the NRC. Before casks are loaded, inspectors with specific knowledge of ISFSI operations assess the adequacy of a "dry run" by the licensee; they then observe all initial cask loadings.

The on-site resident inspectors or region-based inspectors may observe later cask loadings, and the regional offices also perform periodic inspections of routine ISFSI operations.

### What Keeps Fuel Cool in Dry Casks?

The fuel is cooled by natural airflow around the cask. Fuel is often moved to dry cask storage after several years in spent fuel pools, so the heat given off by the fuel has significantly decreased.

## QUESTIONS AND ANSWERS – WASTE CONFIDENCE AND FUTURE PLANS

### How Long Is Spent Fuel Allowed to Be Stored in a Pool or Cask?

NRC regulations do not specify a maximum time for storing spent fuel in pool or cask. The agency's "waste confidence decision" expresses the Commission's confidence that the fuel can be stored safely in either pool or cask for at least 60 years beyond the licensed life of any reactor without significant environmental effects. At current licensing terms (40 years of initial reactor operation plus 20 of extended operation), that would amount to at least 120 years of safe storage.

However, it is important to note that this does not mean NRC "allows" or "permits" storage for that period.

Dry casks are licensed or certified for 20 years, with possible renewals of up to 40 years.

This shorter licensing term means the casks are reviewed and inspected, and the NRC ensures the licensee has an adequate aging management program to maintain the facility.

## The Most Recent Waste Confidence Findings Say That Fuel Can Be Stored Safely for 60 Years Beyond the Reactor's Licensed Life.

**Does This Mean Fuel Will Be Unsafe Starting in 2059 [60 Years After Dresden 1's Original License Ended]?**

**What If the Spent Fuel Pool Runs out of Room Even Before the End of a Reactor License? What Is the NRC Going to Do About This?**

The NRC staff is currently developing an extended storage and transportation (EST) regulatory program. One aspect of this program is a safety and environmental analysis to support long-term (up to 300 years) storage and handling of spent fuel, as well as associated updates to the "waste confidence" rulemaking.

This analysis will include an Environmental Impact Statement (EIS) on the environmental impacts of extended storage of fuel. The 300-year timeframe is appropriate for characterizing and predicting aging effects and aging management issues for EST. The staff plans to consider a variety of cask technologies, storage scenarios, handling activities, site characteristics, and aging phenomena—a complex assessment that relies on multiple supporting technical analyses.

Any revisions to the waste confidence rulemaking, however, would not be an "approval" for waste to be stored longer than before—we do that through the licensing and certification of ISFSIs and casks. More information on the staff's plan can be found in SECY-11-0029.

## Does the Waste Confidence Decision Mean That a Particular Cask Is Safe?

Not specifically.

When the NRC issues of certificates and licenses for specific dry cask storage systems, the staff makes a determination that the designs provide reasonable assurance that the waste will be stored safely for the term of the license or certificate.

The Commission's Waste Confidence Decision is a generic action where the Commission found reasonable assurance that the waste from the nation's nuclear facilities can be stored safely and with minimal environmental impacts until a repository becomes available.

## The Waste-Confidence Revision Seems Like a Long-Term Effort. What Is the NRC Doing to Improve Safety of Spent Fuel Storage Now?

The NRC staff is currently reviewing its processes to identify near-term ways to improve efficiency and effectiveness in licensing, inspection, and enforcement. We expect to identify enhancements to the certification and licensing of storage casks, to the integration of inspection and licensing, and to our internal procedures and guidance. More information on the staff's plans can be found in COMSECY-10-0007.

## The NRC Is Reviewing Applications for New Nuclear Power Plants. What Is the Environmental Impact of All That Extra Fuel?

Continued use and potential growth of nuclear power is expected to increase the amount of waste in storage. This increased amount of spent fuel affects the environmental impacts to be assessed by the NRC staff, such as the need for larger storage capacities. In the staff's plan to develop an environmental impact statement for longer-term spent fuel storage, a preliminary scoping assumption is that nuclear power grows at a "medium" rate (as defined by the Department of Energy), in which nuclear power continues to supply about 20 percent of US electricity production.

## Spent Fuel Storage in Pools and Dry Casks

# Questions and Answers – Security

## What About Security? How Do You Know Terrorists Won't Use All of This Waste Against Us?

For spent fuel, as with reactors, the NRC sets security requirements and licensees are responsible for providing the protection. We constantly remain aware of the capabilities of potential adversaries and threats to facilities, material, and activities, and we focus on physically protecting and controlling spent fuel to prevent sabotage, theft, and diversion. Some key features of these protection programs include intrusion detection, assessment of alarms, response to intrusions, and offsite assistance when necessary. Over the last 20 years, there have been no radiation releases that have affected the public. There have also been no known or suspected attempts to sabotage spent fuel casks or storage facilities. The NRC responded to the terrorist attacks on September 11, 2001, by promptly requiring security enhancements for spent fuel storage, both in spent fuel pools and dry casks.

# Questions and Answers – Emergency Planning

## What Emergency Plans Are Required for Spent Fuel Storage Facilities at Nuclear Power Plants Undergoing Decommissioning or Sites That Have Completed Decommissioning?

Decommissioning reactors continue to be subject to the NRC's emergency planning requirements. For some period of time after the licensee ceases reactor operations, offsite emergency planning will be maintained. This period of time depends on when the reactor was last critical as well as site-specific considerations. Offsite emergency planning may be eliminated when the fuel has been removed from the reactor and placed in the spent fuel pool, and sufficient time has elapsed, such that there are no longer any postulated accidents that would result in offsite dose consequences large enough to require offsite emergency planning. There would be no requirement to maintain offsite systems to warn the public. Onsite emergency plans will be required for both the spent fuel pool and the Independent Spent Fuel Storage Installations, but offsite plans will not be required. If, however, an operating plant is located at the same site as the decommissioning plant, the emergency preparedness plans will still be in effect for the operating plant.

Although offsite emergency planning at a decommissioned site may no longer be required, licensees maintain offsite contacts since any emergency declaration requires notification of state and local officials as well as the NRC. In addition, due to the typically reduced staffs at a decommissioning facility they may rely even more on offsite assistance for fire, security, medical or other emergencies. These reduced EP requirements would remain in effect as long as fuel is onsite.

*(Note: This general description also applies to emergency planning for specifically licensed ISFSIs; those requirements are spelled out in detail in 10 CFR 72.32.)*

# INDEX

## #

9/11, vii, 3, 40

## A

access, 33, 81, 106, 120
accounting, 12, 62, 63
acid, 60
acidic, 29, 112
acidity, 109, 111
age, 15, 85
agencies, 27, 46, 61, 71, 87, 102, 119
Alaska, 21, 53
annual rate, 17, 60
appropriations, 37, 50, 58, 72, 130
assessment, 29, 36, 38, 40, 84, 98, 114, 118,
   119, 135, 139, 141, 142, 143, 144, 146,
   148, 157, 169, 171
atmosphere, 114
atoms, 10, 60
attitudes, 92, 99, 152
Attorney General, 58
Austria, 56
authorities, 46, 111
authority, 10, 47, 104, 106, 122, 124, 141,
   142

## B

ban, 48, 57
barriers, 49, 79, 93, 107, 108, 109, 110, 112,
   113, 115, 116
base, 89, 137, 144
Belgium, 79, 80, 87, 88, 100, 102, 105, 112,
   115, 117, 123, 127, 129, 130, 155, 156
beneficiaries, 84, 146
benefits, 30, 45, 84, 85, 92, 124, 139, 146,
   148
biosphere, 108, 109, 112, 113, 115
boreholes, 77, 93, 96, 97, 108, 110, 124
breakdown, 90
Britain, 125, 129
bureaucracy, 78, 91, 104
burn, 15, 39

## C

Cabinet, 139
candidates, 81, 120
ceramic, 10, 66, 94
certificate, 170
certification, 169, 170
cesium, 30, 60
CFR, 125, 157, 163, 172
challenges, 39, 78, 83, 103, 105, 131, 132,
   145
chemical, 47, 60, 110

174 Index

chemical characteristics, 60
chemical properties, 110
China, 79, 100, 102, 105, 123, 127, 129, 131, 155
circulation, 15, 31
citizens, 99
cladding, 27, 28, 31, 32, 34, 39, 57, 59, 60, 66
classification, 49
climate, 8, 38, 113
climate change, 8, 38, 113
closure, 16, 29, 34, 84, 146, 151
Code of Federal Regulations, 147, 148, 151, 163
commerce, 65
commercial, viii, 2, 4, 6, 7, 8, 9, 10, 11, 12, 13, 17, 18, 21, 23, 24, 26, 27, 28, 29, 34, 35, 41, 46, 47, 48, 49, 53, 56, 59, 60, 61, 62, 63, 64, 66, 67, 85, 103, 155, 160, 162
community(ies), 6, 8, 45, 76, 82, 83, 84, 87, 88, 92, 95, 97, 100, 101, 103, 116, 121, 124, 125, 126, 127, 128, 129, 130, 131, 132, 133, 135, 136, 138, 139, 140, 142, 143, 144, 145, 146, 157
community support, 130
compaction, 109
compliance, 6, 80, 84, 116, 118, 119, 142, 144, 146
composition, 106, 137
conductivity, 115
configuration, 160, 164
Congress, 1, 2, 7, 8, 20, 37, 45, 46, 47, 49, 52, 55, 56, 57, 64, 65, 66, 69, 71, 73, 85, 104, 105, 106, 114, 119, 122, 126, 129, 130, 135, 140, 152, 154, 163
Congressional Budget Office, 72
consensus, vii, 3, 84, 91, 94, 99, 146, 156
consolidation, 4, 14, 46
constituents, 8, 112, 114
construction, 4, 5, 10, 28, 29, 33, 44, 82, 100, 103, 108, 115, 121, 125, 132, 135, 144, 145, 166
consumers, 79, 105, 156
containers, 95
contaminated water, 31, 34

contamination, 26, 35, 67
controversial, 140
cooling, viii, 2, 4, 5, 10, 12, 13, 15, 30, 31, 32, 33, 39, 41, 42, 43, 46, 160, 161, 164, 166
cooperation, 122
copper, 94, 95, 109, 110, 111, 115, 137
corrosion, 13, 28, 31, 34, 66, 95, 110, 111, 112, 113, 114, 115, 137
cost, 4, 10, 26, 40, 42, 47, 48, 49, 105, 107, 156
Court of Appeals, 58, 63, 68
covering, 16, 155
crystalline, 80, 94, 96, 100, 107, 109, 110, 112, 116, 121, 129, 157
culture, 89, 92
current limit, 39
cycles, 50

## D

damages, 9, 58, 160
database, 162
decay, 8, 30, 32, 162
decision makers, 45
decision-making process, 86, 87, 141
deficit, 89, 106
degradation, 31, 34, 115, 137
degradation mechanism, 34
denial, 38
Department of Energy (DOE), viii, 2, 4, 5, 6, 7, 8, 12, 15, 17, 18, 20, 21, 23, 24, 26, 27, 28, 29, 32, 34, 35, 36, 37, 39, 44, 45, 46, 47, 52, 53, 55, 57, 58, 60, 61, 62, 63, 64, 65, 66, 67, 69, 70, 71, 72, 74, 77, 89, 102, 104, 106, 113, 114, 115, 119, 121, 123, 126, 127, 130, 135, 137, 144, 148, 149, 152, 153, 154, 155, 156, 157, 162, 163, 170
Department of Justice, 58, 59
Department of Labor, 71
Department of Transportation, 71
deposits, 105
depth, 13, 82, 95, 96, 121, 134
detectable, 34

Index     175

detection, 165, 171
diffusion, 112
directors, 47
disaster, viii, 2, 40
discharges, 60
dispersion, 111
disposition, 9, 29, 48, 154, 156
distribution, 23
District of Columbia, 58, 63, 68
diversity, 137
DOI, 150
draft, 3, 44, 71, 86, 118, 122, 156
drying, 34, 39

## E

earthquakes, 160
economic incentives, 49
economics, 48
effluent, 67
EIS, 57, 64, 148, 169
election, 125, 131
electricity, 10, 11, 60, 79, 84, 85, 105, 106,
    146, 156, 170
e-mail, 66
emergency, viii, 2, 31, 38, 160, 162, 165,
    167, 171, 172
emergency planning, viii, 2, 38, 167, 171,
    172
emergency preparedness, 172
emergency response, 31, 160
employment, 125
energy, 9, 10, 45, 49, 55, 64, 65, 71, 84, 85,
    99, 102, 103, 146, 164
energy supply, 9
enforcement, 37, 170
engineering, 83
England, 121, 124
environment, 30, 40, 41, 44, 77, 79, 93, 98,
    99, 107, 108, 110, 114, 156, 160, 162
environmental effects, 168
environmental impact, 36, 58, 114, 135,
    136, 137, 139, 155, 161, 169, 170
environmental impact statement, 36, 114,
    170

Environmental Protection Agency (EPA),
    46, 109, 119, 149
environments, 79, 83, 107, 112, 114, 146
equilibrium, 148
equipment, 160, 164, 166
equity, ix, 76, 86, 99, 138
EST, 169
Europe, 56
European Commission, 86
evaporation, 114
evidence, 8, 35, 56, 90, 92, 156
exclusion, 137
execution, 136
exercise, 82, 121, 125
expenditures, 104
expertise, 9
exposure, 15, 30, 80, 110, 116, 117
extraction, 113
extreme weather events, 38

## F

federal facilities, 21, 53
federal government, 5, 8, 9, 47, 64, 94, 102,
    155, 157
Federal Register, 58, 66, 67, 68, 69, 149,
    152
filters, 81, 82, 120, 124, 125, 126, 129, 130,
    134
filtration, 40
financial, 4, 30, 167
Finland, 56, 57, 74, 79, 83, 84, 87, 89, 100,
    102, 103, 105, 115, 120, 123, 124, 126,
    127, 129, 131, 135, 138, 141, 145, 146,
    149, 155
fires, 166
fission, 8, 10, 11, 29, 30, 59, 60, 162, 166
flaws, 111
flexibility, 90, 118, 126
flooding, 164
force, 37, 43
formation, 79, 94, 96, 107, 108, 109, 110,
    112, 121, 123, 124, 130
formula, 105
fractures, 108, 109, 137

fragments, 21, 53
France, 56, 57, 73, 74, 77, 79, 80, 83, 84, 88, 89, 91, 95, 100, 102, 103, 104, 105, 106, 112, 115, 117, 118, 121, 123, 124, 127, 129, 131, 138, 142, 145, 146, 147, 154, 155, 156
fuel management, 5, 30, 32, 48
funding, 5, 37, 39, 49, 84, 105, 146, 155
funds, 79, 105, 106, 107

## G

General Accounting Office (GAO), 58, 65, 67
geology, 79, 81, 93, 98, 107, 120, 124
Georgia, 19, 26, 51
Germany, 33, 57, 61, 66, 77, 78, 79, 80, 100, 102, 104, 105, 115, 117, 120, 121, 123, 125, 127, 129, 131, 142, 149, 155, 156, 157
governance, 148
government revenues, 79, 105, 107
governments, 132, 138, 157
grants, 143
graphite, 61
groundwater, 31, 34, 35, 67, 109, 110, 112, 113, 115, 137, 156
growth, 9, 49, 170
guidance, 15, 34, 35, 124, 134, 144, 170
guidelines, 71, 125, 137

## H

half-life, 60
Hawaii, 21, 53
hazards, 30, 31, 164
healing, 108
health, 8, 30, 35, 71, 80, 81, 117, 119, 136
health effects, 80, 117
heat removal, 65, 163
height, 33
helium, 34
high-level radioactive waste (HLW), vii, ix, 3, 7, 29, 57, 67, 74, 76, 77, 78, 79, 80,

81, 83, 85, 86, 87, 88, 90, 91, 92, 93, 94, 95, 97, 98, 99, 100, 101, 102, 106, 107, 108, 109, 111, 113, 115, 116, 117, 118, 120, 124, 138, 143, 145, 150, 151, 152, 154, 155, 157
history, 70, 76, 86, 89
host, 8, 45, 76, 79, 81, 84, 88, 89, 92, 94, 95, 96, 107, 108, 111, 112, 113, 116, 120, 121, 122, 124, 127, 128, 130, 131, 135, 138, 139, 143, 144, 146, 157
House, 5, 31, 37, 55, 56, 58, 65, 68, 154
human, 30, 77, 93, 99, 117, 156
human health, 30, 77, 93, 99
hybrid, 78, 102
hydrogen, 31, 35, 42, 60, 66, 108
hydrogen atoms, 60

## I

ID, 27, 63, 72
identification, 37, 157
idiosyncratic, 102
image, 11
impact assessment, 138, 152
improvements, 33, 37, 43
independence, 144
individuals, 106
industry, 9, 18, 30, 31, 32, 35, 36, 42, 49, 76, 84, 87, 89, 90, 103, 104, 146, 159, 163, 165
infrastructure, 124, 135
inspections, 31, 32, 35, 43, 164, 168
inspectors, 32, 164, 168
institutions, viii, 2, 78, 92, 101, 104
integration, 170
integrity, 34, 90, 164, 166
interdependence, 156
International Atomic Energy Agency, 56, 62, 74, 86, 149
intrusions, 171
investment, 9, 48, 49
investments, 4, 17, 91
Iowa, 20, 51
iron, 110
irradiation, 59

isolation, 12, 96, 116, 165
isotope, 60
issues, vii, viii, ix, 2, 3, 4, 5, 6, 7, 9, 10, 12, 16, 18, 27, 31, 32, 35, 38, 43, 50, 55, 76, 86, 87, 92, 111, 115, 116, 143, 155, 156, 169, 170

## J

Japan, vii, viii, 2, 3, 5, 8, 13, 31, 33, 37, 56, 57, 68, 69, 70, 74, 78, 79, 91, 100, 102, 104, 105, 123, 124, 125, 127, 134, 139, 152, 155, 157
judiciary, 118
jurisdiction, 5

## K

Korea, 79, 101, 102, 105, 123, 127, 132, 155, 156

## L

laws, 10, 87, 109, 143
lead, 66, 92, 115, 125, 135
leadership, 84, 146
leakage, 31, 67, 165
leaks, 67, 165
learning, 100, 130
legislation, 8, 45, 50, 86, 87, 97, 122, 138, 143, 155
lifetime, 80, 117
light, 10, 11, 13, 34, 37, 48, 119
liquids, 85
litigation, 35
local authorities, 96
local community, 121, 125, 157
local government, 45, 56, 98, 129, 155
Louisiana, 19, 51

## M

magnitude, 42

majority, 120, 134
man, 109
management, vii, ix, 3, 10, 11, 16, 30, 36, 38, 43, 44, 45, 46, 47, 48, 57, 76, 77, 78, 79, 80, 83, 85, 86, 87, 88, 89, 90, 91, 92, 93, 95, 96, 97, 98, 99, 101, 103, 104, 105, 106, 107, 108, 113, 116, 117, 118, 130, 142, 145, 147, 148, 154, 155, 156, 169
Maryland, 19, 21, 51, 53
mass, 10, 23, 24, 51, 59
materials, 12, 21, 27, 28, 29, 34, 39, 40, 46, 48, 53, 55, 57, 59, 60, 61, 64, 66, 93, 111, 116
matter, 77, 91, 106, 108, 122, 130
measurement, 57
media, 71
median, 142
medical, 28, 172
melt, 32
methodology, 84, 118, 137, 141, 144, 146
Mexico, 6, 21, 53, 56, 125, 127, 156
migration, 108
mission(s), 12, 56
Mississippi River, viii, 2, 23
Missouri, 8, 20, 51
models, 8, 47, 142
modifications, 33
modules, 17, 62
Montana, 21, 53
moratorium, 89

## N

National Academy of Sciences(NAS), 4, 5, 8, 33, 34, 40, 41, 64, 65, 66, 69, 72, 74, 77, 85, 90, 94, 95, 108, 116, 137, 150, 151, 156
national policy, 85, 131
National Research Council, 5, 8, 32, 40, 55, 58, 64, 65, 66, 69, 72
natural disaster, 160
natural gas, 63
natural resources, 121
neptunium, 64

178    Index

neutral, 115
next generation, 59
nickel, 114
nodules, 111
Northern Ireland, 149
NSA, 162
nuclear program, 156
Nuclear Regulatory Commission (NRC), v,
    viii, ix, 1, 2, 4, 5, 8, 9, 12, 15, 16, 18, 30,
    31, 32, 33, 34, 35, 36, 37, 39, 40, 41, 42,
    43, 45, 46, 49, 56, 57, 58, 61, 62, 63, 65,
    66, 67, 68, 69, 70, 72, 74, 100, 115, 117,
    119, 123, 130, 144, 151, 152, 157, 159,
    160, 161, 162, 163, 164, 167, 168, 169,
    170, 171, 172
Nuclear Waste Policy Act (NWPA), viii, 1,
    5, 6, 7, 8, 44, 45, 47, 55, 58, 71, 74, 95,
    104, 122, 123, 125, 126, 129, 130, 137,
    140, 147, 154
nuclear weapons, 6, 21, 47, 48, 53, 57, 60,
    61, 63
nuclei, 85
null, 82, 126

## O

Obama Administration, vii, 1, 48
obstacles, 89
officials, 9, 76, 87, 157, 172
Oklahoma, 21, 53
operations, viii, 2, 5, 6, 10, 16, 29, 33, 37,
    44, 53, 57, 69, 100, 101, 103, 156, 163,
    164, 168, 171
opportunities, 76, 87, 144, 145, 155
optimism, 84, 95, 146
optimization, 155
orbit, 98
organic solvents, 60
Organization for Economic Cooperation and
    Development, 74, 149, 150, 151
organizational behavior, 84, 104, 146
organizational learning, 90
overhead costs, 18, 46
overlap, 156
oversight, 38, 43, 104, 111

ownership, 103, 156
oxidation, 109, 111
oxygen, 60, 66, 114

## P

Pacific, 59
parallel, 82, 113, 129, 130, 134, 144
Parliament, 97, 102, 103, 106, 138, 139, 156
participants, 89, 108, 144
pathways, 109
peer review, 84, 113, 146
performance indicator, 35
permeability, 137
permission, 80, 97, 98, 103, 117, 133
permit, 56, 110, 115, 144
plants, viii, 2, 9, 32, 38, 41, 102, 142, 156,
    160, 165
playing, 76, 86
plutonium, 6, 12, 27, 28, 29, 30, 48, 60, 64,
    66, 85, 98, 162
poison, 11
policy, 5, 6, 9, 23, 30, 36, 38, 41, 48, 55, 77,
    82, 90, 93, 102, 124, 129, 154
policy choice, 38
policy issues, 36
political force, ix, 76, 86
political leaders, 83, 146
political opposition, 130
politics, 148
pools, viii, 2, 4, 5, 6, 12, 13, 15, 17, 18, 21,
    23, 25, 26, 28, 31, 32, 33, 34, 35, 36, 37,
    38, 39, 40, 41, 42, 46, 49, 53, 67, 70,
    159, 160, 161, 164, 165, 166, 168, 171
population, 81, 83, 116, 120, 146
population density, 81, 120
power plants, 9, 10, 11, 12, 13, 26, 43, 45,
    88, 102, 103, 125, 135, 156, 159, 161,
    162
precipitation, 114
preparation, 39, 139
preparedness, 38, 43, 62
President, 47, 48, 57, 70, 81, 120, 122, 137,
    140, 149
prevention, viii, 2, 35

principles, 116
private sector, 155
probability, 30, 80, 117
producers, 79, 102, 105, 156
profit, 102
project, vii, 1, 3, 6, 26, 31, 37, 76, 86, 91, 100, 136, 142
proliferation, 9, 48, 50, 60
proposition, 109
protection, 8, 30, 32, 33, 40, 41, 42, 65, 93, 109, 159, 162, 163, 164, 171
prototype, 12
public health, 35, 40, 41, 159
public service, 18, 79, 102, 103
public support, 95, 134
pumps, 15

## Q

qualitative differences, 29
quartz, 94

## R

radiation, 11, 13, 30, 33, 65, 74, 111, 123, 142, 143, 149, 150, 159, 163, 165, 167, 171
radioactive waste, vii, ix, 3, 7, 29, 39, 56, 57, 67, 74, 75, 76, 84, 85, 86, 87, 89, 95, 97, 102, 103, 106, 109, 146, 148, 154, 155
radius, 121
ratification, 82, 134, 137, 140
reactor operators, viii, 1, 31
reading, 70
recommendations, viii, 2, 28, 38, 42, 43, 47, 49, 70, 71, 96, 98, 99, 105, 118, 122, 141, 155, 157
recovery, 29
recurrence, 35
recycling, 48
regulations, 16, 71, 80, 81, 116, 118, 119, 156, 157, 168
regulatory controls, 49

regulatory framework, 37, 49
regulatory oversight, 31, 46, 47
regulatory requirements, 37, 79, 107, 145
reimburse, 156
rejection, 103
reliability, 33, 38
remediation, 26
reprocessing, 5, 6, 7, 9, 12, 17, 21, 26, 28, 29, 34, 47, 48, 49, 53, 55, 57, 60, 61, 62, 63, 66, 85, 94, 97, 98
requirements, viii, 2, 6, 17, 32, 42, 43, 45, 80, 81, 83, 115, 116, 117, 118, 120, 135, 141, 144, 154, 162, 163, 167, 171, 172
researchers, 95
reserves, 106
resolution, 122
resources, 36, 45, 46, 47, 92, 121, 136
response, 32, 33, 35, 41, 78, 87, 103, 119, 137, 171
responsiveness, 90
revenue, 63, 105
risk assessment, 30
risk perception, 92, 138, 148
risk profile, 33
robust design, 165
rods, 10, 13, 28, 34, 60, 159, 164
root, 78, 103, 114
routes, 143
rules, 35, 80, 117, 118, 119, 163
Russia, 57

## S

sabotage, 171
salinity, 137
salt domes, 157
salt formation, 120, 126
science, 12
scope, 27, 29, 38, 42, 47, 60
sea level, 8, 38
security(ies), 5, 9, 18, 40, 41, 42, 44, 45, 46, 106, 160, 162, 163, 167, 171, 172
sediments, 77, 93
self-interest, 89
Senate, 4, 47, 48, 50, 55, 56, 72, 142

# 180            Index

sensitivity, 142
September 11, 32, 33, 44, 160, 171
shape, ix, 15, 76, 86
signs, 90
social acceptance, 81, 87, 120
society, 79, 80, 107, 116
sodium, 13, 61
solubility, 112
solution, ix, 47, 75, 85, 86
sorption, 112, 113, 115
South Dakota, 21, 53
South Korea, 57, 154, 156
Soviet Union, 61
Spain, 79, 87, 101, 102, 105, 123, 128, 132, 155
specific knowledge, 168
spending, 40, 47
spent nuclear fuel (SNF), vii, viii, ix, 1, 2, 3, 4, 5, 6, 7, 8, 9, 10, 11, 12, 13, 14, 15, 16, 17, 18, 19, 20, 21, 22, 23, 24, 25, 26, 27, 28, 29, 30, 31, 32, 33, 34, 35, 36, 37, 38, 39, 40, 41, 42, 43, 44, 45, 46, 47, 48, 49, 51, 53, 54, 55, 56, 57, 58, 59, 60, 61, 62, 63, 64, 65, 66, 67, 69, 70, 71, 74, 76, 77, 78, 79, 80, 81, 83, 85, 86, 87, 88, 90, 91, 92, 93, 94, 95, 97, 98, 99, 100, 101, 102, 103, 106, 107, 108, 109, 110, 111, 112, 113, 114, 116, 117, 118, 120, 124, 135, 138, 143, 145, 150, 154, 155, 157, 159, 160, 167
stability, 84, 146
staffing, 43
state laws, 10
steel, 13, 15, 29, 112, 113, 114, 159, 164, 167
stockholders, 39
storage, vii, viii, ix, 1, 2, 3, 4, 5, 6, 7, 8, 9, 10, 11, 12, 13, 15, 16, 17, 18, 21, 22, 23, 24, 25, 26, 27, 28, 29, 30, 31, 32, 33, 34, 35, 36, 37, 38, 39, 40, 41, 42, 43, 44, 45, 46, 47, 48, 49, 51, 53, 54, 56, 57, 58, 61, 62, 63, 65, 66, 67, 69, 70, 71, 85, 96, 97, 98, 99, 155, 159, 160, 161, 162, 163, 164, 165, 166, 167, 168, 169, 170, 171
stretching, 88

strontium, 30, 60
structure, 3, 13, 32, 33, 38, 42, 47, 59, 103, 123, 124, 144, 165, 166
submarines, 12, 28
substitutes, 35
Sun, 98
surplus, 27
Sweden, 56, 57, 74, 79, 80, 83, 84, 87, 88, 89, 91, 94, 101, 102, 103, 105, 106, 115, 117, 118, 120, 123, 124, 126, 128, 129, 132, 135, 136, 143, 145, 146, 151, 155, 157
Switzerland, 74, 79, 80, 91, 101, 102, 104, 105, 106, 112, 113, 115, 117, 123, 124, 126, 128, 132, 134, 139, 151, 155

## T

tanks, 12, 29, 85, 97
target, 64, 66, 106
Task Force, viii, 2, 5, 35, 37, 42, 43, 61, 67, 68, 69, 70, 89, 153
technical change, 88
technical support, 142
technologies, 4, 7, 39, 40, 48
technology, ix, 9, 18, 30, 36, 39, 41, 48, 49, 60, 62, 76, 86
temperature, 13, 34, 61, 165
tensions, 91
terrorist attack, vii, 3, 33, 41, 160, 171
terrorists, 33, 162
testing, 117, 167
theft, 60, 171
thermal energy, 15
thermodynamics, 109
thorium, 59
threats, 30, 31, 171
TID, 149
time frame, 37, 100, 101
time periods, 118
titanium, 114
tornadoes, 160
toxicology, 30
tracks, 162
training, 12, 43

## Index

transparency, 86, 90
transpiration, 114
transport, 44, 45, 70, 111, 112, 113, 114, 135
transportation, 34, 37, 39, 45, 47, 138, 155, 169
treatment, 29
trust fund, 105, 106
tuff, 80, 108, 113, 115
turbulence, 121

### U

Ukraine, 61
United Nations, 74
universities, 28
updating, 36
uranium, viii, 2, 4, 7, 10, 19, 20, 27, 28, 30, 46, 48, 56, 57, 59, 60, 61, 66, 85, 98, 116, 162
utility costs, 9

### V

variables, 38
venue, 155
veto, 56, 122, 124, 125, 127, 140
voluntarism, 86
Volunteers, 127
vulnerability, 34, 40

### W

Wales, 121, 124
Washington, 20, 21, 35, 52, 53, 55, 60, 61, 62, 65, 72, 137, 148, 149, 150, 151, 152, 153, 154
waste disposal, 6, 45, 108, 155
waste management, 9, 46, 47, 48, 85, 87, 90, 92, 98, 100, 102, 156
water, 8, 10, 11, 13, 15, 23, 31, 32, 33, 34, 35, 41, 48, 60, 61, 65, 108, 112, 113, 114, 115, 121, 159, 161, 164, 165, 166
water chemistry, 13, 34
weapons, 12, 27, 28, 46, 48, 60, 64, 154
web, 56, 67, 69
welding, 111
wet storage pools, viii, 2, 13, 21, 28, 32, 38, 40, 41, 42, 53
Wisconsin, 19, 21, 52, 53
witnesses, 144
workers, 13, 71, 167
World Trade Center, 32
World War I, 60
worldwide, 13

### Z

zirconium, 10, 32, 66